Patenting Genes

Patenting Genes

The Requirement of Industrial Application

Marta Díaz Pozo

PhD, Centre for Commercial Law Studies, Queen Mary University of London, UK

Cheltenham, UK • Northampton, MA, USA

Published by
Edward Elgar Publishing Limited
The Lypiatts
15 Lansdown Road
Cheltenham
Glos GL50 2JA
UK

Edward Elgar Publishing, Inc.
William Pratt House
9 Dewey Court
Northampton
Massachusetts 01060
USA

A catalogue record for this book
is available from the British Library

Library of Congress Control Number: 2016957239

This book is available electronically in the **Elgar**online
Law subject collection
DOI 10.4337/9781786433954

ISBN 978 1 78643 394 7 (cased)
ISBN 978 1 78643 395 4 (eBook)

Typeset by Columns Design XML Ltd, Reading
Printed and bound by CPI Group (UK) Ltd, Croydon, CR0 4YY

Contents

Preface

The major advances in the identification of the human genome that took place from the early 1990s onwards triggered a significant increase in the number of patent applications concerning newly discovered human gene sequences that nevertheless failed to disclose the function of the isolated material, and thus did not meet the patent law requirement of industrial application. To address this issue, the 1998 Directive on the legal protection of biotechnological inventions (Biotech Directive) required patent applicants to disclose the industrial applicability of inventions covering human gene sequences and related proteins at the time of the patent application. Furthermore, the Biotech Directive established functionality-related protection for all types of genetic patents, thus restricting the scope of protection granted to these kinds of inventions to their ability to perform the industrial application disclosed by the applicant. The adoption of those criteria as regards the industrial applicability of gene patents did however contrast with the traditionally vague implementation of this requirement.

This book analyses the implications of the Biotech Directive's approach towards the industrial application of human genes and fragments thereof in respect of three issues: the assessment of the industrial applicability of inventions concerning sequences or partial sequences of human genes; the distinction between discoveries and patentable inventions when the claimed subject matter is human genetic material; and the determination of the scope of protection awarded to patents over genetic information.

It is argued that the Biotech Directive's stringent approach towards the requirement of industrial application can act as an efficient policy option for preventing the grant of patents over human genetic discoveries that are of no practical benefit to society, but also for impeding the issuance of overly broad patents in this field. At the same time, a strict interpretation of this requirement does not necessarily imply that the interests of patent applicants may be systematically overlooked, but it can form the basis of a balanced standard that serves to avoid the rise of undue barriers in the pursuit of research and innovation in this industry.

Acknowledgements

I owe a great debt of gratitude to the Centre for Commercial Law Studies at Queen Mary University of London for believing in this project, as well as to the Herchel Smith research bequest for providing me with the very generous funding to complete this book.

I would also like to extend my sincerest gratitude to Professor Duncan Matthews, for his extremely valuable comments and constant support all along this project, and to Professor Uma Suthersanen, Dr Justine Pila and Dr Herbert Zech for the exchange of the most insightful thoughts.

This book is dedicated to my family and Jose.

Abbreviations

Acad Med	Academic Medicine
AEI	American Enterprise Institute
AIPJ	Australian Intellectual Property Journal
AIPLA	American Intellectual Property Law Association
AIPPI	International Association for the Protection of Intellectual Property
Am Econ Rev	American Economic Review
Am UL Rev	American University Law Review
BC Envtl Aff L Rev	Boston College Environmental Affairs Law Review
BGH	Bundesgerichtshof (German Federal High Court)
BIO	Biotechnology Industry Association
BLAST	Basic Local Alignment Search Tool
BRCA	Breast cancer
BTLJ	Berkeley Technology Law Journal
BUJ Sci & Tech L	Boston University Journal of Science & Technology Law
CAFC	Court of Appeals for the Federal Circuit
Cardozo J Int'l & Comp L	Cardozo Journal of International and Comparative Law
CCD Mass	Circuit Court District Massachusetts
CCPA	Court of Customs and Patent Appeals
CESPRI	Centro di Ricerca sui Processi di Innovazione e Internazionalizzazione (Centre of Research on Innovation and Internationalization)
cDNA	Complementary DNA
Chicago-Kent L Rev	Chicago-Kent Law Review
CIPA	The Chartered Institute of Patent Attorneys
Civ	Civil Division
CJEU	Court of Justice of the European Union
Computer L Rev & Tech J	Computer Law Review & Technology Journal
DDBJ	DNA Database of Japan

Denv UL Rev	Denver University Law Review
DNA	Recombinant deoxyribonucleic acid
EBA	EPO Enlarged Board of Appeal
EBI	European Bioinformatics Institute
EC	European Commission
ECJ	European Court of Justice
Ed	Editor
ED	EPO Examining Division
Edn	Edition
EIPR	European Intellectual Property Review
EMBL	European Molecular Biology Laboratory
Emory LJ	Emory Law Journal
EPC	European Patent Convention
EPO	European Patent Office
EST	Expressed Sequence Tag
EU	European Union
EWCA	Court of Appeal of England and Wales
EWHC	High Court of Justice of England and Wales
Fed Cas	Federal Cases
Fed Cir	Federal Circuit
Fordham Int'l LJ	Fordham International Law Journal
Fordham Intell Prop Media & Ent LJ	Fordham Intellectual Property, Media & Entertainment Law Journal
FSR	Fleet Street Reports
FTC	US Federal Trade Commission
GRUR	Deutsche Vereinigung für gewerblichen Rechtsschutz und Urheberrecht (German Association for the Protection of Intellectual Property)
HGP	Human Genome Project
HGS	Human Genome Sciences
Hous J Health L & Pol'y	Houston Journal of Health Law & Policy
HUGO	Human Genome Organisation
IAM Magazine	Intellectual Asset Management Magazine
Ibid	Ibidem (the same)
IIC	International Review of Intellectual Property and Competition Law
ILM	International Legal Materials
ILP	International Licensing Platform

Ind Int'l & Comp L Rev	Indiana International & Comparative Law Review
Int Arch Occup Environ Health	International Archives of Occupational and Environmental Health
Int'l Intell Prop L & Pol'y	International Intellectual Property Law and Policy
IPC	International Patent Classification
IPQ	Intellectual Property Quarterly
J High Tech L	Journal of High Technology Law
JIBL	Journal of International Biotechnology Law
JIPLP	Journal of Intellectual Property Law and Practice
J Marshall L Rev	John Marshall Law Review
J Pat Off Soc'y	Journal of the Patent Office Society
JPTOS	Journal of the Patent and Trademark Office Society
J Tech L & Pol'y	Journal of Technology Law & Policy
Jurimetrics J	Jurimetrics: The Journal of Law, Science, and Technology
LSG	Law Society's Gazette
MAA	Manufacturers Aircraft Association
Marq Intell Prop L Rev	Marquette Intellectual Property Law Review
Marq L Rev	Marquette Law Review
Mich Telecomm & Tech L Rev	Michigan Telecommunications and Technology Law Review
MLR	The Modern Law Review
MPEG	Moving picture experts group
MSDs	Musculoskeletal disorders
NAPAG	National Academies Policy Advisory Group
Nat Biotech	Nature Biotechnology
Nat Rev Genet	Nature Reviews Genetics
NIH	National Institutes of Health
Nw U L Rev	Northwestern University Law Review
OD	EPO Opposition Division
OECD	Organisation for Economic Co-operation and Development
OLG	Higher Regional Court of Düsseldorf (Oberlandesgericht)
Or L Rev	Oregon Law Review
Pat	Patents Court

PCR	Polymerase chain reaction
PINTO	Patent Information and Transparency On-line
RNA	Ribonucleic acid
RPC	Reports of Patent Cases
Rutgers Computer & Tech L J	Rutgers Computer & Technology Law Journal
SARS	Severe acute respiratory syndrome
SMU L Rev	Southern Methodist University Law Review
SNP	Single Nucleotide Polymorphism
SPC	Supplementary protection certificate
STOA	Scientific Technology Options Assessment
Suffolk UL Rev	Suffolk University Law Review
Supp	Supplement
TBA	EPO Technical Board of Appeal
TNF	Tumour necrosis factor
TRIPS Agreement	Agreement on Trade-Related Aspects of Intellectual Property Rights
U Chi L Rev	University of Chicago Law Review
UCLA JL & Tech	University of California, Los Angeles Journal of Law & Technology
UCL Jurisprudence Review	University College London Jurisprudence Review
UK	United Kingdom
UKIPO	UK Intellectual Property Office
UKSC	UK Supreme Court
UMKC L Rev	University of Missouri-Kansas City Law Review
UN	United Nations
UNESCO	United Nations Educational, Scientific and Cultural Organization
UNSWLJ	University of New South Wales Law Journal
UNTS	United Nations Treaty Series
U Pa L Rev	University of Pennsylvania Law Review
UPOV Convention	International Convention for the Protection of New Varieties of Plants
US	United States
USPQ	US Patents Quarterly
USPTO	US Patent and Trademark Office
Vand J Transnat'l L	Vanderbilt Journal of Transnational Law
Vand L Rev	Vanderbilt Law Review
Vol	Volume

Yale J Health Pol'y L & Ethics	Yale Journal of Health, Policy, Law, and Ethics
WHO	World Health Organization
WIPO	World Intellectual Property Organization
WTO	World Trade Organization

Table of cases

AUSTRALIA

EUROPEAN PATENT OFFICE

EUROPEAN UNION

GERMANY

NETHERLANDS

SPAIN

UNITED KINGDOM

UNITED STATES

Table of legislation

EUROPEAN

INTERNATIONAL

UNITED STATES

1. Introduction

> Nature is the source of all true knowledge. She has her own logic, her own laws, she has no effect without cause nor invention without necessity.
>
> (Leonardo da Vinci)

1.1 BACKGROUND

The term 'biotechnology' was first coined by the Hungarian engineer Károly Ereky in 1919 to refer to the process of producing substances from raw materials with the aid of living organisms.[1] Today, biotechnology is broadly defined as any technological application that uses biological systems, living organisms, or derivatives thereof, to make or modify products or processes for specific use.[2]

The vast potential of biotechnology to improve human and animal health, increase agricultural production and protect the environment has been acknowledged in multiple intergovernmental meetings including the 1992 United Nations (UN) Earth Summit.[3] In fact, over the last few years, the level of innovation in biotech sciences has often been used as an indicator of a country's competitiveness and economic performance. Furthermore, governments worldwide usually consider biotechnology to be a key asset for securing a more competitive and sustainable future. In this regard, promoting innovation in biotechnology currently forms an integral part of the European Union's (EU) strategy towards a knowledge-based economy.

It is commonly known that the origins of biotechnological techniques date back to the ancient times. By 2300 BC Egyptians had already started

---

[1] Robert Bud, 'History of "Biotechnology"' (1989) 337 Nature 10.

[2] UN Convention on Biological Diversity (opened for signature 5 June 1992, entered into force 29 December 1993) 1760 UNTS 79, 143; (1992) 31 ILM 818, art 2.

[3] UN Conference on Environment and Development, Rio de Janeiro, 3–14 June 1992 – the Earth Summit Agreements: Agenda 21, the Rio Declaration on Environment and Development, the Statement of Forest Principles, the UN Framework Convention on Climate Change and the UN Convention on Biological Diversity.

to use microorganisms in the brewing processes for making products such as beer, wine and cheese.[4] However, modern biotechnology, which has had a major impact on the development of today's biotech industry, is mainly based on the scientific advances that have taken place since the nineteenth century.

The science of zymotechnology, which appeared in the late nineteenth century and included all types of industrial fermentation, rather than just brewing, played a crucial role in bridging the gap between old biotechnology and modern genetic engineering.[5] At the same time, Gregor Mendel's experiments on plant and animal breeding made it possible to define the rules of heredity,[6] which represented a tremendous step towards understanding the genetics of organisms. Mendel's theories in combination with the chromosome theory of inheritance by Thomas Morgan constituted the core of classical genetics. The discipline of molecular biology developed in the subsequent decades.

In the twentieth century, growing bonds between biology and engineering led to the development of biotechnology as a science. The 1953 discovery of the deoxyribonucleic acid (DNA) structure and the development of a recombinant DNA technique in 1973 constitute the two key events that set the groundwork for new and revolutionary advances in biotech sciences. Based on Mendel's laws, geneticists began to speculate about the chemical structure and functioning of the gene, and the relationship between genes and proteins, at the start of the twentieth century.

The discovery that DNA is the genetic material of organisms started with Griffith's experiments in 1928, which showed that nonvirulent bacteria strains could be genetically transformed with a substance derived from a heat-killed pathogenic strain.[7] Based on further experimental evidence, Crick and Watson proposed the double-helical structure of DNA, in which two complementary polynucleotide chains (made out of

---

[4] Robert Bud, *The Uses of Life: A History of Biotechnology* (Cambridge University Press 1993) 7.

[5] Bud, 'History of "Biotechnology"' (n 1).

[6] Colin Tudge, *In Mendel's Footnotes: An Introduction to the Science and Technologies of the Nineteenth Century to the Twenty-Second* (Jonathan Cape 2000) 213. Mendel's breeding experiments consisting of genetic crosses between strains of peas with different external characteristics led to understanding the rules of heredity (that heredity is controlled by chromosomes, which are the cellular carriers of genes) from parents to offspring in 1865, see James D Watson (ed), *Molecular Biology of the Gene* (7th edn, Pearson 2014) 6.

[7] Ibid 22.

the primary nucleobases adenine, thymine, guanine and cytosine) are twisted around each other, held together by hydrogen bonds between pairs of bases, to form a regular double helix.[8] In April 1953, the international scientific journal 'Nature' published a communication from the two researchers at the Cavendish Laboratory at the University of Cambridge in England, explaining the double helix molecular structure of DNA.[9] This publication, which was further supported by a second article,[10] proposed the new model of DNA together with some of its genetic implications.

By the autumn of 1953, the 'central dogma' was that chromosomal DNA functions were the template for ribonucleic acid (RNA) molecules, which subsequently moved to the cytoplasm of eukaryotic cells, where they determine the arrangement of amino acids within proteins.[11] Information therefore flows from DNA to RNA and then to protein. In previous decades, investigators had never seen and rarely studied elements such as DNA and proteins.[12] Thus, the scientific foundation established by Crick and Watson allowed a better understanding of previous discoveries, provoking a great transformation of life sciences.

Two decades later, Cohen and Boyer invented a methodology for cutting and splicing DNA segments from different sources, making it possible to transfer genes from one source to the DNA of a different specie. This technology, known as recombinant DNA, set the basis of genetic engineering by allowing the insertion of a fragment of DNA from one cell, for example a human cell, into the genetic sequence of another host cell, for example a bacterium; and then its activation so that the host begins to produce the protein encoded by the inserted DNA sequence. Recombinant DNA has helped investigators to study cancer cells, programmed cell death, antibody production, hormone action and other fundamental biological processes.[13] The practical implications of this

[8] Ibid 24–25. The term 'double helix' became popular with the publication in 1968 of James Watson's book *The Double Helix: A Personal Account of the Discovery of the Structure of DNA* (1st Atheneum paperback edn, Atheneum 1980).

[9] James D Watson and Francis HC Crick, 'Molecular Structure of Nucleic Acids: A Structure for Deoxyribose Nucleic Acid' (1953) 171 Nature 737.

[10] James D Watson and Francis HC Crick, 'Genetical Implications of the Structure of Deoxyribonucleic Acid' (1953) 171 Nature 964.

[11] Watson, *Molecular Biology of the Gene* (n 6) 33.

[12] Eric Vettel, *Biotech: The Countercultural Origins of an Industry* (University of Pennsylvania Press 2006) 176.

[13] Burton E Tropp, *Molecular Biology: Genes to Proteins* (4th edn, Jones & Bartlett Learning 2012) 24.

method include the production of human insulin, which reached the market in 1982 and gradually replaced the insulin extracted from porcine or bovine pancreata,[14] human growth hormones and drugs to treat anaemia, cardiovascular disease and cancer, as well as the development of diagnostic tools to detect a variety of diseases.[15]

Soon after recombinant DNA technology was first developed, the distinction between traditional and modern biotechnology was adopted.[16] 'Traditional biotechnology' refers to the use of living matter in fermentation, like beer brewing or bread making, and plant and animal hybridization. In contrast, 'modern biotechnology' is used to refer to more advanced techniques like manipulation of an organism's genome, the industrial use of DNA cell fusion and tissue engineering among others.

Since the 1990s, biotech research has been principally focused on the act of identifying which stretches of DNA in an organism correspond to which genes, and recombinant products and processes. Moreover, molecular research started to be increasingly focused on commercial applications, which blurred the boundaries between academic (public) and industrial (private) activities.[17] Most recent advances in genetic engineering include protein engineering, nanobiotechnology,[18] DNA shuffling and gene therapy.

With the rapid advancement of genetic engineering techniques, patent protection has become a fundamental asset for companies in this industry.

---

[14] Insulin was first discovered at the University of Toronto in 1921–1922. Eli Lilly and Company prepared insulin in commercial quantities and a patent was granted for the process of obtaining insulin to the University of Toronto in order to control the quality of insulin sold to diabetics, see Michael Bliss, *The Discovery of Insulin* (Macmillan Press 1982) 11–12.

[15] Tropp (n 13) 24–25. Although initial experiments using recombinant DNA raised suspicion, risks from this technique are now considered to be much lower than originally estimated, and more specifically defined, see Lorance L Greenlee, 'Biotechnology Patent Law: Perspective of the First Seventeen Years, Prospective on the Next Seventeen Years' (1991) 68 Denv UL Rev 127.

[16] See Albert Sasson, *Medical Biotechnology: Achievements, Prospects and Perceptions* (UN University Press 2005) 1.

[17] Sally Smith Hughes, 'Making Dollars out of DNA: The First Major Patent in Biotechnology and the Commercialization of Molecular Biology, 1974–1980' (2001) 92 Isis 541. See also Rohini Acharya, Anthony Arundel and Luigi Orsenigo, 'The Evolution of European Biotechnology and Its Future Competitiveness' in Jaqueline Senker (ed), *Biotechnology and Competitive Advantage* (Edward Elgar 1998) 101.

[18] For information about the challenges that nanotechnological inventions may pose to patent law see Herbert Zech, 'Nanotechnology – New Challenges for Patent Law?' (2009) 6 SCRIPT-ed 147.

Patent rights are government-granted limited monopolies attached to inventions that comply with the requirements of novelty, inventiveness (non-obviousness) and industrial application (utility), and are not excluded from patentability.[19] Patents are negative legal rights with potential economic value, which give owners temporary rights to prevent others from using, making or selling a particular invention. Among others, patents can be used for preventing duplication, blocking competitors, securing royalty fees or negotiating licensing contracts.[20] For example, a patent right could be infringed if someone sells or otherwise commercializes the invention without the patent holder's consent, which gives the owner the possibility to enforce his rights in court and obtain compensation for the invention's benefits that have been unlawfully obtained by the infringer, and other costs incurred.

In order to secure continuous research, companies need to somehow guarantee that investors will be able to capture the value of their creations and recover the invested amounts. In this regard, investors usually see patents on the inventions resulting from the financed research as the best indicator that a company is competitive and that the project is achieving its objectives. Thus, having a valuable patent portfolio is very important today for maintaining a competitive market position, especially for biotechnology firms, which require vast amounts of money to carry out research.

During the 1990s and 2000s the gap between policy and life sciences developments became a primary concern in the EU, which led policy makers to focus on adopting different measures to foster research and

---

[19] Under the European Patent Convention there are some exclusions from patentability, either because the subject matter does not constitute an invention, or because although it can amount to an invention, this one is not patentable, see Convention on the Grant of European Patents (European Patent Convention (EPC)) of 5 October 1973, art 52(2) and (3) and art 53.

[20] With regard to the legal nature of patent rights, their analogy with real property is to a certain extent equivocal. Patents are property rights where the term 'property' refers to ownership of knowledge or information. Although patents can be traded or transferred through different types of agreements, knowledge substantially differs from physical things in the sense that knowledge is an inexhaustible good that cannot be the subject of physical control and can be transferred without loss to the transferor and infinitely replicated, see William van Caenegem, *Technology Law and Innovation* (Cambridge University Press 2007) 11–12. In terms of legal validity, the granting of a patent by a country's government does not guarantee the validity of the rights since it can be invalidated in the event of a dispute over the patent's compliance with those patent law requirements that constitute revocation grounds.

innovation in this field. In this regard, patent protection was seen as a crucial element for attracting investors and promoting the successful development of a competitive biotechnology industry in Europe.

The European Patent Convention (EPC) does not contain any express provision excluding genetic inventions from patentability, while Article 27(1) of the Agreement on Trade-Related Aspects of Intellectual Property Rights (TRIPS Agreement)[21] requires that patents are to be available for all fields of technology. However, even though patent laws do not prohibit the patentability of inventions concerning (human) genetic material, the intrinsic characteristics of these types of inventions pose challenges to the application of general patent law rules. For instance, since the launch of the Human Genome Project (HGP) in the 1990s, there has been a significant increase in the number of patent applications concerning human genes and partial sequences, many of which do not easily meet traditional patentability standards. In particular, it has been common to find patent applications for newly discovered, isolated human DNA sequences (and related proteins) for which an industrial application has not yet been identified.

Discussions about the patentability of human genetic inventions were usually focused on questions regarding the novelty, inventiveness or ethical implications of these types of patent applications. However, with the advent of more and more patent claims concerning isolated human genes and gene sequences without a known practical utility in industry, the patent law requirement of industrial application has proved to be a difficult standard to fulfill for these kinds of inventions. In consequence, the criterion of industrial applicability (utility) has become a rather important issue in the debate over the patentability of human genes.

This is especially controversial since the requirement of industrial application is essential to ensure that, where patents are granted for those inventions that make a contribution to the art, they have a real applicability in industry. In the case of gene patents, this requirement serves to guarantee that society is not deprived of using inventions concerning human genetic material, such as diagnostic tests, if the patentee has not been able to indicate what the invention's practical utility in industry is.

The 1998 Biotech Directive attempted to address this question by introducing a number of provisions that give the requirement of industrial application a determinant role in the patenting of human DNA sequences. The aim was to ensure that patents over genes and gene sequences of

[21] Agreement on Trade-Related Aspects of Intellectual Property Rights of 15 April 1994.

human origin contain a real practical value for society that justifies the grant of monopoly rights to such inventions. Moreover, the Biotech Directive did not only raise the industrial applicability standard per se, but further analysis of its text reveals a close link between this requirement and other important provisions in gene patenting, such as the exclusion from patentability of discoveries and the determination of the scope of protection. As a result, the criterion of industrial application itself and in its different dimensions, that is, in connection with other patent law elements, now has a central position in the patenting of genetic material.

1.2 TERMINOLOGY

1.2.1 Biotechnology

This book uses the term 'biotechnology' to refer to the techniques that developed after the discovery of the DNA structure in the 1950s. In particular, it refers to the advances in genetic engineering that form part of what is known as modern biotechnology as opposed to traditional biotechnology's techniques like fermentation.

1.2.2 Genetic Invention, Gene Patent, Genetic Patent, DNA Invention, DNA Patent

The terms 'genetic invention', 'gene patent', 'genetic patent', 'DNA invention' or 'DNA patent' are employed to refer to inventions, or patents over inventions, in the field of genetic engineering. In most cases, the genetic inventions or genetic patents referred to in this book relate to gene sequences, like DNA or amino acid sequences, of human origin. However, in some instances, reference is made to genetic inventions or patents in general, no matter the origin of the genetic material, or to animal or plant related genetic inventions or patents. In all cases, the origin of the genetic material concerned will be specified.

2. Genetic inventions and patent law in Europe

2.1 INTRODUCTION

Although the importance of patents in promoting innovation varies from one industry to another, in the case of the biotechnology industry the specific characteristics of companies make patents a key means for incentivizing research and innovation in this field. Nonetheless, the intrinsic features of biotechnological inventions pose significant challenges to the application of long-standing patent law rules. In Europe, growing interest in promoting innovation in biotechnology and genetics led to the adoption of different rules and policies to facilitate the patentability of biotechnological inventions, including those concerning human genetic information. In particular, the 1998 Biotech Directive[1] set the basis for a new era of patent law and practice regarding the patentability of living matter. This chapter introduces the policy and legal context for the next chapters of this book. After a brief overview of the economic rationale and justifications for the creation of patent systems, the chapter studies the importance of patents in promoting innovation in genetics and then goes on to explore the background and process for the adoption of the 1998 Biotech Directive as well as existing EU policies for the promotion of research and innovation in biotechnology.

2.2 PATENTS AND HUMAN GENETICS

2.2.1 The Economics of the Patent System

The economic rationale and justification for the grant of patent monopolies has been extensively discussed in the patent law and economics literature; therefore this section does not intend to overlap with existing

[1] Directive 98/44/EC of the European Parliament and of the Council of 6 July 1998 on the legal protection of biotechnological inventions [1998] OJ L213/13.

works, but it provides a brief account of the main economic arguments that sustain the creation of patent systems and highlights the historical importance for inventions to have some practical (industrial) value. The concept of 'invention' is similar but different from 'innovation'. Invention relates to ideas as such and precedes the patent while innovation refers to the application and marketing of inventions and thus follows the patent.[2] Innovation can be defined as a change in ideas, practices or objects involving some degree of novelty and success in application. In contrast, 'innovation system' refers to the utilization of innovations through a network of various actors within an institutional framework.[3] Patent rights are believed to have a significant impact on both the level of innovation and the functioning of innovation systems.

The emergence of patents is part of the evolution of market economies and capitalism. In Antiquity and the Middle Ages, although inventors were rewarded by further contracts or promotions, there was no formal protection of inventions and organized secrecy prevailed.[4] The idea that some kind of monopoly over inventions could be used to promote innovation appeared in the Greek city of Sybaris in 500 BC, where the king would grant one-year exclusive rights to chefs who invented a new recipe.[5] Then in the late Middle Ages, different kinds of privileges rewarding inventions were granted in northern Italy in order to attract foreign craftsmen with new abilities.[6] A commonly cited example is the Brunelleschi's patent for a system that would transport marble up the Arno River. The patent was granted by the city of Florence in 1426 and

---

[2] Fritz Machlup, 'An Economic Review of the Patent System' (1958) Study No15 of the Subcommittee of Patents, Trademarks and Copyrights of the Committee on the Judiciary – United States Senate 85th Congress, Second Session, 56.

[3] See Bjorn Asheim, Finn Valentin and Christian Zeller, 'Intellectual Property Rights and Innovation Systems: Issues for Governance in a Global Context' in David Castle (ed), *The Role of Intellectual Property Rights in Biotechnology Innovation* (Edward Elgar 2009) 38.

[4] Dominique Guellec and Bruno van Pottelsberghe de la Potterie, *The Economics of the European Patent System: IP Policy for Innovation and Competition* (Oxford University Press 2007) 15.

[5] Ibid 16.

[6] These privileges gave the lord of the manor or the merchant guilds the right to monitor and regulate trade within a defined area and thus restricted competition by limiting access to the market, see Michael Fysh and others, *The Modern Law of Patents* (2nd edn, LexisNexis Butterworths 2010) 1013.

formalized in the Venice Statute in 1474,[7] which contained all the essential elements of modern patent law, namely:

- the patentability requirements
- the requirement of registration
- basic rules of infringement
- right to license
- limited duration and
- remedies of damages and delivery up.[8]

In particular, the requirement of industrial application, or for an invention to be useful, was an essential part of this very early form of patent law and continued to be a key element in the subsequent evolution and justification of patent systems (see section 3.2.1). Although much more refined, this criterion along with the other provisions was included in the Venice Statute, which constitutes the basis of modern patent systems.

Soon after the Venice Statute was adopted, patents started to be granted in England, France and Germany. The 1624 Statute of Monopolies was the basis for the further development of patent laws in England. By the early nineteenth century, patent laws had been passed in France and Germany, and were then passed in many other countries in continental Europe. This was related to the spread of the industrial revolution and market economies and the emergence of the modern state and the rule of law.[9] In fact, the development of patent laws during the Modern Era in Europe occurred simultaneously with the progressive introduction of new manufacturing processes in European industries. In that context, a primary aim of patent laws was to promote industrial development, which gave the requirement for an invention to be useful in industry a key role in achieving that objective. At the end of the nineteenth century, most advanced countries had a robust system of patent law, so governments started to focus on harmonization. In 1883 the adoption of the Paris Convention set the basis for future steps towards harmonizing intellectual property laws.[10] In respect of patents, European harmonization efforts

---

[7] Harold C Wegner, *Patent Harmonization* (Sweet & Maxwell 1993) 2; JA Goldstein and E Golod, 'Human Genes Patents' (2002) 77 Acad Med 1315; Guellec and van Pottelsberghe de la Potterie (n 4) 16.

[8] Fysh and others (n 6) 1014–1015.

[9] Guellec and van Pottelsberghe de la Potterie (n 4) 21.

[10] Paris Convention for the Protection of Industrial Property of 20 March 1883. The Paris Convention contains substantive provisions regarding all types of industrial property including patents, trademarks, industrial designs, geographical

culminated with the adoption of the Strasbourg Convention in 1963[11] and the subsequent EPC in 1973. Innovation is currently a crucial element in the determination of a country's level of development and competitiveness. The idea that the true wealth of nations mostly lies in their innovative activity began to prevail after World War II, and has been reinforced by the increasing global competition in technological innovation.[12] It is argued that innovation is essential for overcoming today's growing scarcity of resources,[13] and thus establishing mechanisms for promoting investment in innovative research has been a priority for governments in most countries. As regards the economic rationale behind patent systems, patents can be described as an incentive to invest in creating knowledge the practical application of which results in disclosure. Their principal aim is to encourage innovation and investment in innovation, but also dissemination of knowledge throughout the economy.[14] The economic rationale for patent rights is that without such protection, the level of innovation would fall below that socially desirable because imitation might occur so quickly that an inventor may not be able to appropriate sufficient return of his invention, which would discourage him to continue investing in innovative research.[15] Therefore, in order to stimulate inventive activity and the flow of useful inventions, society provides incentives such as patent rights. Throughout history economists have repeatedly acknowledged the role of patents in promoting creativity and technical innovation. In 1776 Adam Smith recognized the importance of patents to safeguard innovation, arguing that the grant of a temporary trade monopoly to a company venturing into a new market, such as the grant of a patent to an inventor, is a way for the state to compensate innovators for 'hazarding a dangerous and expensive

---

indications, trade secrets and utility models, and also provisions against unfair competitive practices. It is administered by the World Intellectual Property Organization (WIPO) and has a total of 176 Contracting Parties as of July 2016.

[11] Convention on the Unification of Certain Points of Substantive Law on Patents for Invention (Strasbourg Convention) of 27 November 1963.

[12] Sigrid Sterckx, 'The Ethics of Patenting – Uneasy Justifications' in Peter Drahos, *Death of Patents* (Lawtext Publishing Limited and Queen Mary Intellectual Property Research Institute 2005) 177.

[13] William van Caenegem, *Technology Law and Innovation* (Cambridge University Press 2007) 5.

[14] Asheim, Valentin and Zeller (n 3) 37.

[15] Erich Kaufer, *The Economics of the Patent System* (Routledge 2001) 19.

experiment, of which the public is afterwards to reap the benefit'.[16] Under this view, the patent system would basically be a means to promote technical, economic and, ultimately, social progress. Further, the thesis that the grant of monopolies promotes innovation is commonly associated with Joseph Schumpeter's analysis on how monopoly conditions may enable and accelerate innovation and growth more effectively than competition in a dynamic capitalist system. According to Schumpeter, this can be achieved through a process of creative destruction[17] in which the incessant creation of new innovations by new firms replaces old firms that provide obsolete goods and services to the market. The economists Machlup and Penrose later distinguished four fundamental arguments to justify the creation of patent rights,[18] each of which has been challenged to some extent in the patent literature. In sum, there is the natural property rights theory, according to which society is morally obligated to recognize and protect a man's property rights over his ideas (resulting from his own work). This approach predominates in Europe and usually relies on John Locke's labour theory to acquire ideological legitimacy as a justification for the grant of patents,[19] as opposed to the American utilitarian view of intellectual property rights as rights granted by governments discretionally.[20] Second, the reward theory or incentive to invent theory, which requires society, as an act of fairness, to reward a man through the grant of monopoly rights for providing society with useful inventions while impeding free use of those inventions. Third, the incentive to innovate theory sustaining that neither invention nor exploitation of invention will be maintained at an optimal level unless inventors and capitalists have assurances that successful ventures will yield profits

[16] Adam Smith, *An Inquiry into the Nature and Causes of the Wealth of Nations* (edited by Edwin Cannan in 1904, first published by W Strahan and T Cadell in 1776) 712.

[17] See Joseph Schumpeter, *Capitalism, Socialism and Democracy* (5th edn transferred to digital printing, Routledge 2005) 81–86. For the opposite view that competition favours innovation see Kenneth J Arrow, 'Economic Welfare and the Allocation of Resources for Inventions' in Richard R Nelson (ed), *The Rate and Direction of Inventive Activity: Economic and Social Factors* (Princeton University Press 1962).

[18] Fritz Machlup and Edith Penrose, 'The Patent Controversy in the Nineteenth Century' (1950) 10 The Journal of Economic History 1.

[19] As Peter Drahos noted, 'Locke on property has a totemic status' for the justification of patent rights, see Peter Drahos, *A Philosophy of Intellectual Property* (Dartmouth 1996) 41, 48–49.

[20] Robert P Merges, *Justifying Intellectual Property* (Harvard University Press 2011) 94.

which will make their efforts, and risking their money, worthwhile. And finally, the incentive to disclose theory, which assumes that in the absence of patent protection inventors would keep their inventions secret preventing the public from gaining the full benefit of new knowledge and leading to wasteful duplicative research in order to prevent imitation from competitors.

The public disclosure of patent information is usually perceived as the most immediate benefit from the patent system since it allows competitors to gain easy access to the technical knowledge pertaining to a patent application and exploit its value. Because patent information can be used as an indicator of the state of the art, patent disclosure also helps to avoid the coexistence of overlapping research projects. Moreover, well-defined patents allow competitors to undertake more transparent business negotiations with the patent owner,[21] which ultimately facilitates technology transfer.

In the late 1990s Mazzoneli and Nelson proposed four broad theories about the purposes of patents,[22] namely:

- the 'invention motivation theory', according to which the anticipation of patents motivates inventors to create useful inventions
- 'induce commercialization theory', according to which patents help to obtain the necessary investments to develop and commercialize inventions
- 'information disclosure theory' according to which patents are awarded to individuals who disclose their inventions in return and finally
- 'exploration control theory', which supports the idea that patents enable the orderly exploration of a broad prospect.

---

[21] Ove Grandstrand, *The Economics and Management of Intellectual Property: Towards Intellectual Capitalism* (Edward Elgar 1999) 77. Disclosure was not always a key component of the patent system. It was only in the twentieth century that it became central in Europe. In England for example, until the mid-eighteenth century no full disclosure was required, even to the authority in charge of examination. Lord Justice Mansfield wrote in 1778 'you must specify upon record your invention in such a way as shall teach an artist, when your term is out, to make it and to make it as well as you by your directions: for then at the end of the term, the public have the benefit of it', see Guellec and van Pottelsberghe de la Potterie (n 4) 40.

[22] Roberto Mazzoneli and Richard R Nelson, 'The Benefits and Costs of Strong Patent Protection: A Contribution to the Current Debate' (1998) 27 Research Policy 273.

On this basis, the objectives of the patent system could be summarized as follows:

- providing incentives for innovative behaviour
- inducing investment to develop and commercialize technologies
- encouraging the disclosure of information and
- facilitating the transfer of technology.[23]

These theories are not mutually exclusive and they are likely to overlap with each other depending on the characteristics of the particular industry and case.

Despite the numerous economic theories attempting to capture the essence of the role of patents in innovation systems, today there is still not a universally accepted theoretical justification for the grant of patent rights and commentators continue to justify the grant of patents using essentially the same arguments that existed for the creation of the first patent systems. Besides, even though patents can play an important role in promoting the research and development of new products, assuming that patent protection will always improve the incentives to innovate might be misleading in many cases. Empirical evidence has shown a wide diversity across industries in the effectiveness of patents as a means of appropriation (either to prevent duplication or to secure royalty income) and also as a function of firm size.[24] An extensive analysis of the role of patents in innovation goes beyond the scope of this book. However it is important to note that with a few notable exceptions such

---

[23] Richard Y Boadi, 'The Role of IPRs in Biotechnology Innovation: National and International Comparisons' in David Castle (ed), *The Role of Intellectual Property Rights in Biotechnology Innovation* (Edward Elgar 2009) 399.

[24] See Christopher T Taylor and Z Aubrey Silberston, *The Economic Impact of the Patent System: A Study of the British Experience* (Cambridge University Press 1973) 365; Richard C Levin and others, 'Appropriating the Returns from Industrial Research and Development' [1987] 3 Brookings Papers on Economic Activity (Special Issue On Microeconomics) 783; Wesley J Cohen, Richard R Nelson and John P Walsh, 'Protecting Their Intellectual Property Assets: Appropriability Conditions and Why the US Manufacturing Firms Patent (Or Not)' [2000] NBER Working Paper No 7552. See also Mark Shurmer, 'Standardisation: A New Challenge for the Intellectual Property System' in Andrew Webster and Kathryn Packer (eds), *Innovation and the Intellectual Property System* (Kluwer Law International 1996) 49.

as pharmaceuticals and biotechnology,[25] economists have been unable to show a clear causal link between patent rights and increased innovation. In other industries, time-leading, reputational advantage, moving quickly down the learning curve, sales or service efforts have in many cases been regarded as more effective means for appropriating the returns of innovation than patent exclusive monopoly rights.

2.2.2 Patent Rights and Genetic Research

As noted above, the effectiveness of patents in encouraging innovation differs according to the characteristics and needs of each industry. Thus, in order to understand the role of patents in the progress of innovation in gene technologies, it is important to explore the particular relationship between patents and innovation in genetics. In biotech industries, companies possess characteristics that give patents a very important role as an incentive to innovate. They are knowledge-intensive enterprises the activities of which include basic and applied research, as well as systematic development of products and processes. Further, in general patents are most likely to support the growth of knowledge-intensive industries in fields characterized by low ratios of imitation costs, such as in areas with large-scale research projects that result in highly codified knowledge. This is typically the case of science-based industries, such as life sciences, which attempt to use existing knowledge but also advance scientific research and capture the value of the knowledge it creates.[26] For example, in the case of gene-based innovations, the cost of isolating and characterizing a gene and putting it into a commercially viable format can be very expensive, but once the discovery is complete, the genetic information can be easily duplicated.[27] In this regard, patents help to prevent imitators from easily appropriating a competitor's invention.

On the other hand, innovation in biotechnology, especially in the field of genetics, moves at a very high pace and new products are constantly reaching the market, thus firms require access to large amounts of capital

---

[25] Robert W Hahn, 'An Overview of the Economics of Intellectual Property Protection' in Robert W Hahn (ed), *Intellectual Property Rights in Frontier Industries: Software and Biotechnology* (AEI-Brookings Joint Center for Regulatory Studies 2005) 11.

[26] Grandstrand (n 21) 238; Gary P Pisano, 'Can Science Be a Business? Lessons from Biotech' (2006) 84 Harvard Business Review 114; Asheim, Valentin and Zeller (n 3) 43.

[27] Lee Bendekgey and Diana Hamlet-Cox D, 'Gene Patents and Innovation' (2002) 77 Acad Med 1373.

to finance their projects and be at the same level as their competitors. In fact, the state of competitiveness of this industry is strongly equated with the level of capital raised.[28] Moreover, the activities involved in this business require specific qualifications and capabilities. Thus in order to be successful, companies need to assemble a highly educated workforce with valuable research experience or high-level university training, which also entails significant costs to firms in this industry.

In Europe, most biotech companies are small, research-led enterprises with a project-based form of organization that obtain funds through pre-seed investment, or investment in highly promising ideas at a very early stage, which adds value to the project and helps to attract investors.[29] Firms then receive further funding from investors and industrial collaborators based on the evaluation of the project's achievements.[30] In order to recoup the invested amounts, companies protect their inventions with patents, so that they can license or sell the patented products with high margins, and also ensure that competitors cannot duplicate them.[31] Thus potential investors see patent applications as a guarantee that their investments will be recovered,[32] which helps to overcome the general investment risk that goes along with technological uncertainty. Patents have also provided universities and public research

---

[28] Gary P Pisano, *Science Business: The Promise, the Reality, and the Future of Biotech* (Harvard Business School Press 2006) 162.

[29] Albert Sasson, *Medical Biotechnology: Achievements, Prospects and Perceptions* (UN University Press 2005) 66–67.

[30] Steven Casper, *Creating Silicon Valley in Europe: Public Policy towards New Technology Industries* (Oxford University Press 2007) 18.

[31] Biotechnology Industry Association (BIO), 'BIO's Testimony on Patent Reform to Maximize Innovation in the Biotechnology Industry' (25 March 1999) <http://www.bio.org/advocacy/letters/bios-testimony-patent-reform-maximize-innovation-biotechnology-industry> accessed 4 September 2014; UK BioIndustry Association, 'Letter to the Guardian regarding gene patenting' (23 November 2000) <http://www.bioindustry.org/newsandresources/bia-news/letter-to-the-guardian-regarding-gene-patenting/> accessed 13 February 2013; BIO, 'Statement on US Supreme Court review of isolated DNA patents' (13 June 2013) <http://www.bio.org/media/press-release/statement-us-supreme-court-review-isolated-dna-patents> accessed 5 August 2014. Evidence has been found that there is a positive correlation between stock price, when a patent is filed and issued, and research and development expenditures, see David H Austin, 'Estimating Patent Value and Rivalry Effects: An Event Study of Biotechnology Patents' (1994) Resources for the Future (Washington, DC) Discussion Paper 94-36.

[32] Graham Dutfield, *Intellectual Property Rights and the Life Science Industries: A 20th Century History* (Ashgate 2003) 153; Philip W Grubb and Peter R

institutions, which play a key role in the evolution of biotech and genetic sciences, with further incentives to perform their research and to frame it in a way that would facilitate downstream applications.[33] Furthermore, as many biotech companies are not capable of going downstream, doing clinical tests or manufacturing drugs, having a competitive patent portfolio that they can license or sell allows them to continue in the market. Hence the impact of patents in promoting research is not confined to the benefit of strong, diversified biotech companies, but it extends to public institutions and smaller businesses in this industry. Since the 1970s the contribution of advanced technologies to the economic performance of most developed countries has grown significantly. Currently the competitiveness of a nation heavily depends on its level of innovation in certain fields like information technologies and life sciences.[34] Therefore, given the importance of patents in the progress of biotechnology in general and genetic research in particular, in order to stimulate innovation and economic growth, countries started to encourage the use of patent rights in such promising industrial areas. Promoting the patentability of inventions in this field was seen as a key policy measure for the further expansion of the industry. The main difference between biological inventions, such as inventions concerning genetic substances, and any other type of invention lies in the fact that the starting material is living matter, which triggers several legal and moral concerns. For instance, it has been argued that biotechnological inventions amount to mere discoveries that are not capable of meeting the traditional patent law requirements, or that biotech patents are contrary to generally accepted standards of morality.[35] In particular, the patentability of biotech inventions concerning isolated human genetic material has been highly controversial due to the intrinsic characteristics and the sensitive nature of this

---

Thomsen, *Patents for Chemicals, Pharmaceuticals, and Biotechnology: Fundamentals of Global Law, Practice, and Strategy* (Oxford University Press 2010) 275.

[33] Guellec and van Pottelsberghe de la Potterie (n 4) 123.

[34] Dutfield (n 32) 17.

[35] For example, under the slogan patenting lives, it has been argued that patenting living matter equates to owning human life. For a criticism of the 'patenting life' argument, see Stephen Crespi, 'Biotechnology Patenting: The Wicked Animal Must Defend Itself' (1995) 17 EIPR 431. For a discussion about the appropriateness of applying the property discourse to the allocation of rights to human biological materials, see E Richard Gold, *Body Parts: Property Rights and the Ownership of Human Biological Materials* (Georgetown University Press 1996). For a discussion about the ownership implications of the discovery of DNA and mapping of the human genome as academic, universal, personal,

type of subject matter, which falls within the grey area between public and private research and raises questions regarding the privatization of human heritage or the potential problems of access to research results, among others.

As regards the application of the classical patentability criteria, since the 1980s patent laws and practices have generally had serious difficulties keeping up with the rapid scientific progress in the biotechnology field,[36] especially in human genetics. A particular example is the case of the requirement of industrial application. As stated earlier, major advances in the discovery of the human genome immediately gave rise to a growing number of patent applications over human genetic sequences. However, many patent applicants filed their claims before a plausible use of the discovered sequences in industry had been identified, and thus failed to meet the patent law requirement of industrial application. A 1985 report from the Organisation for Economic Co-operation and Development (OECD) observed that most countries wanted biotechnology to adapt to patent law requirements and not vice versa, and that the idea of adaptation or relaxation of the ground rules of patent law and practice for a particular technology and for inventions of a certain type was clearly controversial.[37] Proposals such as improving patent quality through raising the patentability standards, reducing uncertainty in patent law decisions[38] and controlling costs[39] are frequently presented as technical solutions to the general difficulties in the patent system. However none of these claims have had a major impact; instead, patent laws and practices have been the subject of significant reviews and amendments to facilitate the grant of patents concerning genetic material. The increasing number of expectations that the biotech industry was raising led to a desire to establish biotechnology patent law as a specialty area of legal practice that would offer specific and flexible standards for patenting biotech inventions, including those concerning human genetic material. In this

---

national, intellectual and taxable property, see Jasper A Bovenberg, *Property Rights in Blood, Genes and Data: Naturally Yours?* (Martinus Nijhoff Publishers 2006).

[36] Grubb and Thomsen (n 32) 275.

[37] Friedrich-Karl Beier, R Stephen Crespi and Joseph Straus, *Biotechnology and Patent Protection: An International Review* (OECD Publishing 1985) 41, 44 and 46.

[38] Kaufer (n 15) 50.

[39] Christopher May, 'On the Border: Biotechnology, the Scope of Intellectual Property and the Dissemination of Scientific Benefits' in David Castle (ed), *The Role of Intellectual Property Rights in Biotechnology Innovation* (Edward Elgar 2009) 268–269.

context, the 1998 Biotech Directive set the grounds for a new era in biotechnology patenting in Europe. However the Biotech Directive not only recognized the patentability of biological inventions, which smoothed the path to obtaining patents on living material, but it also addressed important concerns regarding the interpretation of patent law rules within the context of human genetic inventions such as the implementation of the requirement of industrial applicability.

2.3 THE EC DIRECTIVE ON THE PROTECTION OF BIOTECHNOLOGICAL INVENTIONS

2.3.1 The First Proposal for a Directive in Context

Over the last few decades, biotechnological advances and their applications in medicine, industry and agriculture have been the focus of increasing attention in the EU. Both the 1983 Communication to the Council entitled 'Biotechnology in the Community'[40] and the 1985 Commission White Paper on Completing the Internal Market (the 1985 White Paper) emphasized the need to improve the regulatory environment surrounding biotechnology, particularly the system of intellectual property rights, in order to facilitate the creation of a single market and promote the competitiveness of the European industry.[41] Subsequently, the 1986 Single European Act introduced a new Article 8a of the 1957 Treaty establishing the European Economic Community,[42] providing for the Community to adopt the decisions and measures necessary to progressively establish the internal market and implement the programme described in the 1985 White Paper.

Also during the 1980s, important international events concerning genetic research and innovation took place such as the 1980 US Supreme Court decision in *Diamond v Chakrabarty*,[43] which confirmed that genetically modified organisms could be patented and formally opened the door for the possibility of patenting living matter; and important advances in recombinant DNA techniques which led, for example, to the

[40] European Commission (EC) Communication to the Council 'Biotechnology in the Community' [COM(83) 672 final/2] – Annex of October 1983.

[41] EC White Paper from the Commission to the European Council on Completing the Internal Market of 14 June 1985 [COM(85) 310 final] 39.

[42] Single European Act signed at Luxembourg on 17 February 1986, OJ L169/1, art 8a.

[43] *Diamond v Chakrabarty*, 447 US 303 (1980).

discovery and patenting of human insulin. By the end of the decade, the feeling among European policy makers was that biotechnology was a key scientific field for the progress of innovation and competitiveness of the Community that needed to be promoted.

In as early as 1883 the Paris Convention defined the scope of patentable subject matter in a manner that included at least part of biotechnological inventions. It stated that the words 'industrial property' used in the Convention were to be understood in the broadest sense, thus including not only products of industry in the strict sense, but also agricultural and mineral products, and all manufactured or natural products, for example:

- wines
- grain
- tobacco leaf
- fruit
- cattle
- minerals
- mineral waters
- beer
- flowers and
- flour.[44]

In 1988 the principal international agreements concerning the protection of biotechnological inventions were the 1961 International Convention for the Protection of New Varieties of Plants (UPOV Convention),[45] the 1963 Strasbourg Convention and the 1973 EPC. The UPOV Convention offers a sui generis form of intellectual property protection specifically adapted to protect plant varieties resulting from plant breeding processes with the aim of encouraging breeders to develop new varieties of plants. By contrast, like the Paris Convention, Article 3 of the Strasbourg Convention follows a flexible approach towards patentable subject matter.

---

[44] Paris Convention (n 10) art 1(3). See also WIPO National Seminar on Intellectual Property, *The International Protection of Industrial Property: From the Paris Convention to the TRIPs Agreement* (Lecture prepared by Professor Michael Blakeney, Cairo 17–19 February 2003) 4. Nonetheless, although the Paris Convention introduced a number of important provisions, the Biotech Directive took its basis on subsequent international agreements on intellectual property.

[45] International Convention for the Protection of New Varieties of Plants (UPOV Convention) of 2 December 1961.

Moreover, although Article 2b of the Strasbourg Convention explicitly excludes plant and animal varieties from patentability, it does not preclude the patentability of inventions concerning any other type of living matter such as human genetic material.

Further developments in harmonizing patent laws resulted in the 1973 EPC and the 1975 Convention for the European Patent for the Common Market (Community Patent Convention), which never entered into force.[46] Both conventions reflected the basic principles of the UPOV and the Strasbourg Convention with regard to biotechnological inventions but failed to provide solid provisions for the patentability of living matter. Within this context, the European Commission (the Commission) presented to the Council of Ministers (the Council) a first proposal for a Directive on the legal protection of biotechnological inventions (First Directive Proposal) in October 1988.[47] The First Directive Proposal emphasized the importance of biotechnological research and development and considered the multiple challenges that the patent system encounters when applied to biotechnology. In this regard, although previous international efforts to improve legal protection for biotechnological inventions had not had significant impact, the First Directive Proposal relied on the existing legislative framework in biotechnology patenting with the objective of improving it. In fact, the text of the First Directive Proposal included a fairly detailed assessment of the existing laws on biotech patenting, particularly the EPC.

In view of the existing legal framework, the Commission noted that the legal situation suffered from deficiencies and discrepancies in statutory provisions, regulations and their interpretation as well as a general shortage of case law, which might ultimately impede the proper functioning of the internal market.[48] For example, the First Directive Proposal pointed out that the EPC did not consider the scientific developments that took place over the ten-year period previous to its approval; neither had it been amended to consider changes in technology after it came into force.[49]

The proposal for the creation of a Directive on the legal protection of biotechnological inventions therefore built on the idea that the existing

---

[46] Agreement related to Community Patents signed at Luxembourg on 15 December 1989 (89/695/EEC) [1989] OJ L401/1.

[47] Commission of the European Communities proposal for a Council Directive on the legal protection of biotechnological inventions (1988) [COM(88) 496 final – SYN 159] OJ C10/3.

[48] Ibid, points 14 and 15.

[49] Ibid, point 22.

legal system had become outdated due to scientific and technological developments, and was thus unable to satisfy the needs of the European industry. In essence, the principal objective of the First Directive Proposal was to establish harmonized, clear and improved standards on the protection of biotechnological inventions, including the patentability of living matter (such as human genetic material), and on the effects of the exclusions from patentability of plant and animal varieties, scope of protection and sufficiency of disclosure. It was believed that the adoption of a Biotech Directive would help to attract internal and external investments in biotech research, which would in turn facilitate the development of the European biotechnology industry, the trade of biotechnological products and the establishment of a common market in this field.

2.3.2 A Ten-Year Approval Procedure

In order to adopt domestic laws in the EU, a cooperative procedure between the Commission, the Council and the European Parliament (the Parliament) must be followed. The procedural route for the adoption of the Biotech Directive was the (ordinary) co-decision procedure of Article 189b of the Treaty of Maastricht.[50] Under the co-decision procedure, the Commission presents a proposal to the Parliament and the Council, which must be approved by a qualified majority of the Council and an absolute majority of the Parliament before it is adopted.

Incidentally, while the Parliament was considering the First Directive Proposal, the European Patent Office (EPO) agreed to grant a patent for the at that time controversial Harvard Onco-mouse patent application, which had been previously denied on the grounds that animals were excluded from patentability according to Article 53(b) of the EPC. However, in 1990 the EPO Technical Board of Appeal (TBA) ruled that only animal varieties as such were excluded and that the invention was not contrary to public order or morality. The TBA found that the process claim on the Harvard Onco-mouse was not an essentially biological process according to a balancing test that weighted the animal's suffering

[50] Soon after the First Directive Proposal, the Maastricht Treaty came into force introducing the co-decision procedure in Article 189b, see Treaty on European Union of 29 July 1992 (Treaty of Maastricht) [1992] OJ C191/1, art 189b (which corresponds with Article 251 of the Consolidated version of the Treaty establishing the European Community of 24 December 2002 [2002] OJ C325/33, now Article 294 of the Consolidated version of the Treaty on the Functioning of the European Union of 30 March 2010 [2010] OJ C83/49).

against benefits for humanity.[51] This decision showed that the EPO approach to Article 53 was to interpret exceptions and limitations to patent rights very narrowly, which caused controversy in the Parliament.[52]

By contrast, in October 1990 the United States (US) Department of Energy and National Institutes of Health (NIH) launched the HGP, an international scientific research project with the objective of determining the sequence of chemical base pairs in DNA, and identifying and mapping the approximately 20 000–25 000 genes of the human genome from both physical and functional perspectives. The three billion dollars publicly funded project received the support from geneticists in the UK, France, Germany, Japan, China and India, and was finally completed in 2003.[53] The HGP noted the importance of genetic research and the need to promote continuous innovation in this field. In this context, while patents on biotechnological inventions relating to the human body raised concerns about access to knowledge and ethical questions, there was the view that precluding the patentability of biotech inventions could have a negative impact on the development of the industry, which is highly dependent on patent monetization.

In October 1992 the Parliament voted to approve the First Directive Proposal subject to its amendments,[54] and in December of the same year the Commission approved an amended proposal (1992 Amended Proposal), which accepted half of the 44 amendments proposed by the Parliament. However, the 1992 Amended Proposal was rejected by the Parliament in March 1995, mainly due to ethical issues. In the Parliament's opinion, the proposal removed too many restrictions on the

---

[51] *Onco-mouse/HARVARD* (T 0019/90) [1990] (EPO (TBA)) point 5.

[52] See Janice McCoy, 'Patenting Life in the European Community: The Proposed Directive on the Legal Protection for Biotechnological Inventions' (1993) 4 Fordham Intell Prop Media & Ent LJ 501.

[53] A few years later, in 1998 the US biotech company Celera Genomics launched a similar project, although privately funded. Unlike the HGP, Celera announced that it would seek patent protection for a significant part of their findings and soon after that, the firm filed preliminary patent applications on 6500 genes. The high number of patent applications and the firm's reluctance to allow free distribution and scientific use of their findings triggered worldwide reactions such as the joint statement issued by US President Clinton and UK Prime Minister Blair in March 2000 declaring that the human genome sequence could not be patented and should be made freely available to researchers everywhere.

[54] Amended Proposal for a Council Directive on the legal protection of biotechnological inventions (1992) [COM (92) 589 final – SYN 159] OJ C44/36.

patenting of life forms,[55] notwithstanding the fact that the TRIPS Agreement also gave members the possibility of excluding from patentability inventions the commercial exploitation of which is deemed to be contrary to public order or morality, and that the moral question was already considered in Article 53(a) of the EPC. The Commission on the contrary considered that these questions were already sufficiently addressed in other legal instruments and thus focused on technical and harmonization objectives.

After the Parliament's rejection of the amended version of the First Directive Proposal, the Commission presented a new draft proposal in December 1995 (1996 Amended Proposal), which was reviewed by six different Committees.[56] After renewed opposition in the Parliament and the adoption of 65 of the 66 proposed amendments, a modified version was adopted in October 1997 (1997 Amended Proposal).[57] The new text introduced changes in response to the Parliament's concerns, but also incorporated a new particular provision (Article 5(3)) aimed at providing a means to address the harmful effects that patents over isolated human DNA fragments of uncertain function might have on the progress of innovation in the biotech industry.[58] The new version was accepted by the Parliament and approved by the Council by a qualified majority

---

[55] The different competing interests in the patentability of living matter resulted in three main forces competing for public support in the debate over the adoption of the Directive: the 'Greens' (term generally used to refer to those who raise moral or ethical objections to the patenting of living matter usually based on environmentalist arguments), family farmers (primarily concerned with the patentability of genetically modified plants, seeds and related products) and the various biotech trade associations (which strongly supported the approval of the Biotech Directive), see David G Scalise and Daniel Nugent, 'Patenting Living Matter in the European Community: Diriment of the Draft Directive' (1992) 16 Fordham Int'l LJ 990.

[56] The new proposal was sent to the Council in January 1996, see Commission proposal for a European Parliament and Council Directive on the legal protection of biotechnological inventions (1996) [COM(95) 661 final] OJ C296/4. The six Committees that reviewed the new proposal were Agriculture and Rural Development; Development and Co-operation; Economic and Monetary Affairs and Industrial Policy; Environment, Public Health and Consumer Protection; Research, Technological Development and Energy; and Legal Affairs and Citizens' Rights (which was the leader).

[57] Amended Proposal for a European Parliament and Council Directive on the legal protection of biotechnological inventions (1997) [COM(97) 446 final].

[58] The background and reasoning for adopting this requirement are analysed in Chapter 4.

(although the Netherlands voted against it) in November 1997.[59] The proposal was subsequently agreed to as the Common Position of the Council,[60] and finally on 12 May 1998 the Parliament formally gave its approval.[61] The text of the Biotech Directive, consisting of a 65-paragraph preamble and 18 articles, was formally adopted on 16 June 1998 (Council Decision of 16 June 1998) and published in the Official Journal in August 1998.

After ten years of discussions, the Biotech Directive finally became part of European Community law. Member States had two years from the date of entry into force (30 July 1998) to ensure that their national laws were consistent with the new text. However, by 2003 only six Member States had implemented the Biotech Directive. A number of countries failed to implement the new legislation (Austria, Belgium, France,

---

[59] On 19 October 1998, the Kingdom of the Netherlands (supported by Italy and Norway) brought an action for annulment of the Biotech Directive. The Council and the European Parliament were the defending parties and the Commission intervened in support of the Directive. The application brought by the Netherlands put forward six pleas concerning the incorrect legal basis for the Directive: breach of the principle of subsidiarity, breach of the principle of legal certainty, breach of obligations in international law, breach of the fundamental right of respect for human dignity and breach of procedural rules in the adoption of the Commission's proposal. In addition, the Netherlands submitted an interim application to the President of the Court of Justice of the European Community aimed at postponing the implementation of Biotech Directive. According to the Kingdom of the Netherlands, transposal would have had serious and irreversible consequences that could not be rectified in future. By an injunction of 25 July 2000, the President of the Court rejected that application. The pleadings took place before the Court of Justice on 13 February 2001. The conclusions of Advocate-General Jacobs were delivered on 14 June 2001 and recommended that the action for annulment be dismissed. The judgment of the Court of 9 October 2001 upheld the conclusions of the Advocate-General and dismissed the action, see C-377/98 *Kingdom of the Netherlands v European Parliament and Council of the European Union* [2001] ECR I-07079; EC Report from the Commission to the Council and the European Parliament: 'Development and Implications of Patent Law in the Field of Biotechnology and Genetic Engineering' [COM(2002) 545 final] (First Commission Report 2002) 8. For further comments on the CJEU's decision in response to the Netherlands' action for annulment of the Directive see Sebastian Moore, 'Challenge to the Biotechnology Directive' (2002) 24 EIPR 149; Robert N Weekes, 'Challenging the Biotechnology Directive' [2004] European Business Law Review 325.

[60] Council Common Position of 26 February 1998 [1998] OJ C 110/17.

[61] Decision of the European Parliament of 12 May 1998 [1998] OJ C167/5.

Germany,[62] Italy, Luxembourg, the Netherlands, Portugal and Sweden) until required to do so by the courts in 2004–2005. Implementation of the Biotech Directive into national law across all Member States was not completed until 2007.

The reasons why the majority of EU Member States resisted implementing the Biotech Directive differed from country to country, however in a workshop held by the OECD in June 2001, some concerns were identified, such as:

- potential dependency resulting from (broad) DNA patents
- reluctance of researcher to enter fields with already patented genes
- monopolistic genetic testing practices
- royalty stacking and
- explosion of legal disputes.[63]

Put in a different way, the main concerns expressed around patenting biotechnology mainly related to patents concerning human genes and their effect on the progress of research.

With regard to the compatibility of the Biotech Directive with the TRIPS Agreement and the EPC, when dealing with the Netherlands' action for annulment the Court of Justice of the European Union (CJEU) did not consider itself competent to assess the validity of the Biotech Directive with regard to the EPC, in that the European Community is not a party to it;[64] and likewise, the Court declined its competence with regard to the compatibility of the Biotech Directive with the TRIPS Agreement, to which the European Community is a party for those aspects for which it is competent, in view of the fact that that Agreement is based on a principle of reciprocal and mutually advantageous arrangements.[65] Moreover, with regard to biotechnology, Article 27 of the TRIPS Agreement takes a non-discriminatory approach towards this kind of invention. The Biotech Directive's approach towards biotechnology relies

[62] Even though Germany, together with the UK, was the most progressive country with patenting rules on biotechnology that paralleled those of the US and Japan, it failed to implement the Directive within the established time framework, see Scalise and Nugent (n 55).

[63] Joseph Straus, 'Product Patents on Human DNA Sequences' in F Scott Kieff (ed), *Perspectives on Properties of the Human Genome Project* (Elsevier/Academic Press 2003) 67.

[64] *Kingdom of the Netherlands v European Parliament and Council of the European Union* (n 59) point 53.

[65] Ibid, point 53.

on Article 27 and further confirms that the fact that the subject matter of an invention is living material does not impede access to patent protection.

2.4 CURRENT EU POLICY TOWARDS PATENTING BIOTECHNOLOGY

The 1985 Commission White Paper on Completing the Internal Market and the subsequent 1998 Biotech Directive established the basis for a new phase in the EU biotech industry. Then in March 2000, the European Council met in Lisbon to assess the Union's strengths and weaknesses in order to establish the basis for a structural reform in the EU (the Lisbon Strategy). The Lisbon Strategy set long-term growth and social targets for the next decade, 'to become the most competitive and dynamic knowledge-based economy in the world, capable of sustainable economic growth with more and better jobs and greater social cohesion'.[66]

The means to achieve this goal included preparing the transition to a knowledge-based economy and society by creating better policies for research and development and stepping up the process of structural reform for competitiveness and innovation. To foster economic growth, employment and social cohesion, under the new knowledge-based economy innovation and ideas would be adequately rewarded, particularly through patent protection.

The Lisbon Strategy established the context for the next steps in the advance of biotechnology in Europe, which was considered to be a frontier technology with the potential to enhance Europe's competitiveness, by setting general guidelines that triggered a series of interdependent reforms. In this regard, the 2002 Commission Communication 'Life Sciences and Biotechnology – A Strategy for Europe' proposed a comprehensive action plan for addressing the challenges posed by the biotechnology revolution in the knowledge-based economy.[67] It urged the development of pro-active policies to exploit the opportunities from

---

[66] Presidency Conclusions, Lisbon European Council, 23–24 March 2000. See also Presidency Conclusions, Stockholm European Council, 23–24 March 2001.

[67] EC Communication from the Commission to the Council, the European Parliament, the Economic and Social Committee and the Committee of the Regions of 23 January 2002: 'Life Sciences and Biotechnology – A Strategy for Europe' (2002) [COM(2002) 27 final] OJ C55/3, accepted in March 2002 by the Barcelona Declaration.

the positive potential of life sciences and biotech, ensure proper governance and address Europe's international responsibilities. The Communication proposed 30 actions including the creation of a strong, harmonized and affordable intellectual property protection system. To achieve these goals, Member States needed to transpose the Biotech Directive's provisions into their national laws (Action 5).[68]

A mid-term assessment report of the Lisbon Strategy's achievements revealed that slow and limited progress had been made[69] and stated the need to re-double efforts in order to meet the strategy's goals. With respect to biotechnology, new strategic actions were adopted to promote the European biotech industry and the initial action plan on life sciences was refocused on specific and priority sectorial issues.[70] In particular, intellectual property goals were re-directed towards more precise and defined objectives. Patent protection came to the forefront of the strategies to foster innovation, competitiveness and knowledge transfer in life sciences.[71] In this sense it was believed that effectively implementing the Biotech Directive would lead to the creation of an accessible and efficient patent system, which would in turn promote biotech innovation.[72]

The policies described above show that EU policy makers consider patents to be a key means for boosting biotechnology innovation and research. The Commission has continuously stressed the need to remove obstacles and reward innovation in promising industries such as biotechnology. Thus one might expect that patent laws will continue to protect biotechnological and genetic inventions to the same (or further) extent in the coming decades.

---

[68] Ibid 36.

[69] EC, *Facing the Challenge: The Lisbon Strategy for Growth and Employment – Report from the High Level Group chaired by Wim Kok* (European Commission 2004).

[70] After completion of the mid-term assessment of the Lisbon Strategy, the 2002 Commission Communication 'Life Sciences and Biotechnology – A Strategy for Europe' was revised in 2005 and 2007, see EC Report from the Commission to the European Parliament, the Council, the Committee of the Regions and the European Economic and Social Committee of 29 June 2005 'Life Sciences and Biotechnology – A Strategy for Europe' Third progress report and future orientations [COM(2005) 286 final]); EC Communication from the Commission to the Council, the European Parliament, the European Economic and Social Committee and the Committee of the Regions on the mid-term review of the Strategy on Life Sciences and Biotechnology [COM(2007) 175 final].

[71] Communication mid-term review 2007 (n 70) 8.

[72] Ibid 10.

2.5 CONCLUDING REMARKS

In knowledge-intensive industries like genetics, patents play a key role in helping companies to recoup the high amounts of capital invested to finance their projects. In this regard, after a long negotiation process between the Commission, the Council and the Parliament, the approval of the Biotech Directive established a new direction in the patenting of biotechnological inventions. The purpose of the Biotech Directive was to harmonize national patent laws regarding biotechnology throughout Europe, but also address some of the problematic issues that arise from the patenting of living matter, such as the conditions for the patenting of human genetic material, the increasing number of patent applications over isolated human gene sequences of unknown function or the scope of patent protection for genetic inventions.

The chapters that follow focus on the current approach towards the industrial applicability of inventions concerning human DNA sequences and its implications for the interpretation of other important patent law provisions regarding the patenting of human genetic information, such as the exclusion from patentability of discoveries and the scope of patent protection awarded to genetic inventions.

3. The European requirement of industrial application

3.1 INTRODUCTION

Novelty, inventive step and industrial application constitute the three essential conditions that an invention has to fulfill in order to receive patent protection. History shows that the requirement of industrial application has always been a central aspect in an invention's definition in modern patent systems. However, more recent interpretations of this requirement have set a very low threshold that has relegated industrial application to an almost inexistent patent law standard.

This chapter first studies the historical development of the European requirement of industrial application. The chapter then examines the rationale and objectives of this requirement and the meaning and interpretation of Article 57 of the EPC and related provisions on industrial applicability. Finally, the chapter discusses existing arguments in favour and against adopting a strict approach towards the assessment of an invention's practical utility.

3.2 THE DEVELOPMENT OF THE REQUIREMENT OF INDUSTRIAL APPLICATION

3.2.1 Historical Background

The principal aim of the patent system is to encourage innovation and social progress through granting exclusive monopolies over most sorts of inventions. To this end, rather than offering a definition of 'patentable invention', European patent law provides a set of requirements that an invention has to fulfill in order to be worthy of the exceptional rights that a patent monopoly confers.[1] Thus, if an invention is not excluded from

[1] Convention on the Grant of European Patents (European Patent Convention (EPC)) of 5 October 1973, arts 52(2) and (3), and art 53.

patentability and meets the standards established by the law, it should then be patentable. The three patentability requirements of the current European patent system are the result of a long process of legal and policy debates over the question of what kinds of objects deserve to be the subject of discretionary government-granted monopoly rights. Hence there is the novelty requirement to ensure that society is not prevented from using things that are already part of the 'state of the art'. There is also the requirement of inventive step, which impedes inventors from obtaining patents for things the creation of which is obvious to society. And finally, there is an industrial application requirement to guarantee that patented inventions can be utilized in industry; or, in other words, that the public can use those inventions and benefit from them.

The history of European patent law reveals that the notion of industrial application has had a principal role in the development of today's patent system. Historically the function of the patent system was to advance the industrial arts, and thus social progress, by supporting new industries and industrial development.[2] In particular, the events that took place during the period in which modern European patent laws were enacted influenced the inclusion of the criterion of industrial application in patent law regimes. In the seventeenth century a major scientific revolution took place in Europe during which the works of authors like Francis Bacon, Galileo Galilei, René Descartes or William Harvey on applied science with utilitarian ends, as well as their ideas regarding physical laws, built the basis for western ideas of social and industrial progress.[3] In that context, the adoption of the standard of industrial application in European patent laws was overall coherent with the prevailing objective at that time of fostering the development of tangible objects with a real utility in practice.

Notwithstanding the influence of the Early Modern period in the conception of industrial applicability as a criterion for the granting of patent monopolies, the notion of utility can also be found in the first forms of patent systems known in Europe. Patents began to be used by governments in later Medieval and early Renaissance Europe primarily to encourage the transfer and disclosure of more advanced foreign industrial practices through the emigration of skilled artisans.[4] Under the Tyrolean

---

[2] Justine Pila, *The Requirement for an Invention in Patent Law* (Oxford University Press 2010) 8.

[3] Ibid 8.

[4] Paul A David, 'The Evolution of Intellectual Property Institutions and the Panda's Thumb' (Meetings of the International Economic Association, Moscow, August 1992).

system in the Middle Ages, which was the basis of the Venetian system and the word invention, patents were granted or revoked by the state, depending on what was deemed to be useful.[5] Novelty and inventiveness were only occasionally investigated while the main requirement was practical applicability.[6] In contrast, the Venetian Statute of 1474 referred to the inventor of 'any new and ingenious device, not previously made within our jurisdiction'. However the preamble to the statute's text emphasized that the social and economic utility of such monumental legal innovation both depended on, and corresponded to, the practical utility of the inventions themselves.[7] Soon after the Venetian Statute was adopted, the notion of patent law, including the requisite for an invention of being useful, spread from Venice to Germany, France and England.

In Germany, inventions started to be patented about ten years after the adoption of the Venetian Statute.[8] The criterion of practical usefulness as a pre-requisite for patent protection continued to be present in subsequent patent practice and a few centuries later, Article 1(1) of the first German Patent Act (Reichspatentgesetz) in 1877 stated that an invention had to be 'gewerblich verwertbar' (susceptible to industrial application) in order to be patentable.[9] Following that provision patents were granted for inventions that were 'novel and allow for industrial use'.[10] Industrial applicability was viewed as the central criterion for delimiting the scope of patentable inventions based on whether the claimed subject matter was part of trade and industry in the patent law meaning of the term.[11] Furthermore, in connection to the requirement of industrial application, in the nineteenth century Germany developed the requirement of 'technical advance' (against the state of art). This doctrine was unique to

---

[5] Harold C Wegner, *Patent Harmonization* (Sweet & Maxwell 1993) 5.

[6] Ibid.

[7] Christopher Wadlow, 'Utility and Industrial Applicability' in Toshiko Takenaka (ed), *Patent Law and Theory* (Edward Elgar 2008) 359. See also Jeremy Phillips, 'The English Patent as a Reward for Invention: The Importation of an Idea' (1982) 3 The Journal of Legal History 71.

[8] Historical evidence suggests that the Venetian patent idea was imported by travelling German businessmen and immigrant Venetian glassmakers, see John F Duffy, 'Harmony and Diversity in Global Patent Law' (2002) 17 BTLJ 685.

[9] See Frithjof E Muller and Harold C Wegner, 'The 1976 German Patent Law' (1977) 59 J Pat Off Soc'y 91; Wadlow (n 7) 364.

[10] Maximilian Wilhelm Haedicke and Henrik Timmann (eds), *Patent Law: A Handbook on European and German Patent Law* (CH Beck, Hart and Nomos 2014) 125.

[11] Ibid.

German and Swiss law and established that a patentable invention had to demonstrate its technical superiority over what was previously known in the art. However, although this approach may be regarded as having taken the concept of utility to its logical utilitarian conclusion, the standard of 'technical advance' was abolished under the new law after the adoption of the EPC.[12] In fact, as in other European countries, the substantive significance of the requirement of industrial applicability has gradually decreased over the past century[13] with the significant exception of the biotechnology field.

Similarly, the use of patent grants to encourage the development of new industrial practices as an instrument of mercantilist policy spread in France during the mid-sixteenth century.[14] The French patent system, which developed in the wake of the Venetian system, provided that the Royal Academy of Science shall examine all machines for which privileges are solicited from His Majesty and certify whether they are new and useful.[15] The law of 1791 formally established a patent system in France and continued the practices of the 'Ancien Régime' under which inventors received royal privileges freeing them to exploit their inventions outside the confines of existing guild controls.[16] As happened in Germany, the policy justification to grant patent monopolies in France was precisely to encourage inventors to put their inventions into practice in public, so that society could benefit from new technologies. Nonetheless, the specific requirement of industrial application was added to the new law on patents of July 1844 according to which only the 'new industrial products', the 'new ways or new applications of known means for obtaining a result or an industrial product' could be patented, provided that the new invention, discovery or application is described by the applicant.[17] Hence since its origins dating back to the early Tyrolean system, the requirement of industrial application was further reinforced during the Early Modern European period and ultimately given a principal position in modern continental patent law regimes.

In the UK, the first definition of patentable invention appeared in Section 6 of the 1623 British Statute of Monopolies. The Statute granted letters patent (open letters) and grants of privilege in respect of any 'manner of new manufacture', and the courts solved the question of what

---

[12] Muller and Wegner (n 9); Wadlow (n 7) 358 and 365.
[13] Haedicke and Timmann (n 10) 125–126.
[14] David (n 4).
[15] See Wegner (n 5) 5–6.
[16] David (n 4).
[17] French Law on Patents of 5 July 1844, art 2 and art 5.

constituted an invention by asking whether the invention was a manner of manufacture for which patents could be granted under the Statute of Monopolies. In this regard, the term 'manner of manufacture' revealed a close relationship between what should be a patentable invention and the practical arts; it suggested that patentable inventions were those that actually had an application in practice.[18] Furthermore, the verb 'to invent' at that time carried very extensive connotations that were mainly related to industrial development objectives, such as bringing into use, find, establish or institute manufacture.[19] Nonetheless, the idea of utility itself first appeared in English law in a 1624 discourse about the Statute of Monopolies by Sir Edward Coke, who stated that 'in every such new manufacture that deserves a privilege, there must be "urgens necessitas" and "evidens utilitas"'.[20] 'Evidens utilitas' had no expression in the statute but since early times the courts accepted that inventions had to show both novelty and utility.[21]

The case law of the nineteenth and part of the twentieth centuries on the concept of manner of manufacture further showed that the idea of practical application (utility) was present in the courts' understanding of what constituted a patentable invention within the meaning of the Statute.[22] In 1932, utility was included in the 1907 British Patents and Designs Acts as one of the 16 grounds upon which a patent could be revoked.[23] Subsequently, the 1949 consolidating Act clarified that every claim in a patent application had to be useful.[24] Therefore, like continental patent regimes, UK patent law has maintained the standard of industrial application as a fundamental requirement for the granting of patents.

---

[18] Ng-Loy Wee Loon, 'Patenting of Genes – A Closer Look at the Concepts of Utility and Industrial Applicability' [2003] IIC 393; Sivaramjani Thambisetty, 'Legal Transplants in Patent Law: Why Utility is the New Industrial Applicability' Law, Society and Economy Working Papers 6/2008 (2009) 49 Jurimetrics J 155.

[19] David (n 4).

[20] Edward Coke, *The Third Part of the Institutes of the Law of England: Concerning High Treason, and Other Pleas of the Crown, and Criminal Causes* (4th edn, first published by Andrew Crooke and others in 1669) 184.

[21] Wee Loon (n 18).

[22] Pila (n 2) 31–34, 72–75.

[23] UK Patents and Designs Act of 1907 as amended in 1932.

[24] UK Patents Act of 1949, section 32(g) stating that a patent may be revoked by the court if the invention, so far as is claimed in any claim of the complete specification, is not useful.

In fact, when discussions about harmonization of patent laws started in the mid-1800s, once patent laws of various sorts were in place in most major European countries, including the British Empire, France and Germany,[25] European domestic laws varied widely;[26] however, all of them had some bearing on the criterion of industrial application.

3.2.2 The Formulation of Article 57 of the EPC

In response to growing interests in obtaining patent protection in foreign markets, national governments worldwide started to negotiate bilateral treaties for the protection of industrial property.[27] However, these treaties did not prove to be sufficiently effective in Europe so countries started to seek further unification of their patent laws.

The Council of Europe of 1949 represented a leading step towards the development of a harmonized or unified patent system in Europe. The main underlying reasons behind the idea of harmonization were that unifying national laws would contribute to improved relations between the people of Europe and, in more economic terms, that the fragmented national approach to protect technical inventions was contrary to good sense and rational economic behaviour.[28] Discussions for the harmonization of the substantive patent laws of the different countries followed over the next two decades and finally led to the adoption of the Strasbourg Convention of 1963[29] and the subsequent EPC of 1973.

The patent law requirement of industrial application set out in Articles 52(1) and 57 of the EPC has its origins in the Strasbourg Convention of 1963, which represented the first step in the creation of a European patent. Developed under the Paris Convention's provision for the making of special agreements between Union States for the protection of industrial property,[30] the Strasbourg Convention sought to harmonize certain points of substantive patent law for promoting European industry

---

[25] Alexander James Stack, *International Patent Law: Cooperation, Harmonization, and an Institutional Analysis of WIPO and the WTO* (Edward Elgar 2011) 65.

[26] Ibid.

[27] Ibid 66.

[28] Friedrich-Karl Beier, 'The European Patent System' (1981) 14 Vand J Transnat'l L 1.

[29] The Strasbourg Convention harmonized certain aspects of patent law that provided the template for the future negotiations of the EPC.

[30] Paris Convention for the Protection of Industrial Property of 20 March 1883, art 1(3).

and establishing a common market in Europe,[31] and contributing to the future creation of an international patent system.[32] An early draft of the agreement already described the patentability criteria by providing that the object of the application for a European patent must belong to the domain of technology, must be new and must have a quantum of invention.[33] Subsequent work on substantive patent law focused, among others, on the question of the practical character of patentable inventions.

The original proposal to the Council for the creation of a European patent system was submitted by the French senator Longchambon in 1949.[34] Following this, a Committee of Patent Experts (the Committee) was formed to study the proposal, with representatives from all the Member States of the European Council, and in 1953 a comparative study of national European laws was undertaken by the Secretariat-General of the Committee.[35] Although the Longchambon proposal was finally dismissed, the Committee continued the preparatory works for the creation of a patent system in Europe. They carried out a national questionnaire,[36] but given the imprecision of and diversity between national laws, in 1955 the Committee resolved to unify only those aspects of substantive law that were necessary to create a European system for granting patents.[37] Hence the Committee identified a few patentability conditions as potential subjects of harmonization, namely:

---

[31] NS Sreenivasulu and CB Raju, *Biotechnology and Patent Law: Patenting Living Beings* (Manupatra 2008) 38.

[32] Convention on the Unification of Certain Points of Substantive Law on Patents for Invention (Strasbourg Convention) of 27 November 1963, Preamble; Pila (n 2) 115. For further reading about harmonization of patent law see Stack (n 25).

[33] Wee Loon (n18).

[34] Longchambon's plan for harmonizing European patent law was submitted to the Consultative Assembly of the Council of Europe on 6 September 1949, see Consultative Assembly Documents, 1 Session (1949), Council of Europe Doc No 75. See also Justine Pila, 'Article 52(2) of the Convention for the Grant of European Patents: What Did the Framers Intend? A Study of the Travaux Preparatoires' (2005) 36 IIC 755.

[35] Committee of Experts, *Comparative Study of Substantive Law in force in the Countries Represented on the Committee of Experts on patents presented by the Secretariat-General (Mr Gajac)*, EXP/Brev (53) 18 (7 November 1953), see Pila (n 2) 134.

[36] Committee of Experts, *Questionnaire drawn up by the Bureau of the Committee of Experts on Patents*, CM/12 (52) 149 (17 November 1952) 4–5, see Pila (n 2) 132.

[37] Pila (n 2) 134.

- industrial character
- novelty
- technical progress
- creative effort and
- influence of prior patent rights.

Each of those questions was considered to be an essential element for the constitution of a common patent system in Europe.

With regard to industrial character, the discussions of the Committee about the requirement of industrial application shed light on the importance of this question for defining an invention. The Committee remarked that, apart from novelty, the industrial characteristic was the only one that was required of a patentable invention by all the national regulations.[38] For example in Germany, Austria, Belgium, Denmark, Greece and the Netherlands inventions were required to be 'capable of industrial application', in France and Turkey the law required inventions to 'arise from any kind of industry', and in the UK and Ireland in order to be patentable inventions needed to 'be manners of manufactures'.[39] Then in a 1960 Memorandum,[40] the Rapporteur-General noted a 'certain uniformity of outlook' in respect of the concept of industrial character. In his opinion, the comments showed that the patent systems existing in Europe were characterized by the 'extreme terseness' of their legal provisions on industrial character, the usual formula being:

- 'inventions which could be applied for industrial purposes'
- the existence of a large body of case law built up over many years
- the significant similarity between the structure of case law in different countries over the concept of industrial character and
- the existence of certain divergences, mainly in the sector of agriculture, derived largely from considerations of public interest.[41]

---

[38] Ng-Loy Wee Loon (n 18). See also Richard Beetz, Dieter Behrens and Wolfgang Dost, *European Patent Law: Practicing under the European Patent Convention* (EPC) (Heymanns 1979) 10.

[39] See Ng-Loy Wee Loon (n 18); Beetz, Behrens and Dost (n 38) 10.

[40] Committee of Experts, *Memorandum on the unification of legislation (Item 4 of the Agenda for the meeting of 28th November 1960) by Rapporteur-General Mr. Finniss*, EXP/Brev (60) 7 (28 November 1960), cited in Pila (n 2) 140.

[41] Pila (n 2) 140.

Hence, despite existing differences between domestic laws, a comparative analysis of the relevant provisions revealed that the requirement of industrial application ultimately built upon similar basis in all European countries.

Harmonization on the question of industrial application was finally achieved by using the wording 'susceptible of industrial application', which was the one employed in most European patent laws, and included inventions in the field of agriculture within the scope of the requirement. The final text of the Strasbourg Convention employs the word 'susceptible' rather than 'capable', although the latter has also been used in some European countries in their definition of industrial application. In fact, the election of the term 'susceptible' was purely a matter of semantics and as the EPO Guidelines for Examination later explained,[42] both terms have the same meaning in the context of industrial application and can be used interchangeably.

The negotiations finally resulted in the formulation of Article 1 of the Strasbourg Convention stating that patents shall be granted for any inventions that are susceptible of industrial application, which are new and which involve an inventive step. With regard to the industrial application requirement, Article 3 then specified that an invention should be considered as susceptible of industrial application if it can be made or used in any kind of industry including agriculture.[43] In this respect, the explanatory memorandum of the preparatory documents for the draft of the Strasbourg Convention stated that Article 3 is concerned with the industrial character of the invention and is to be understood in the wide sense of Article 1(3) of the Paris Convention.[44] These provisions were reproduced a decade later when they were incorporated in the text of the EPC.

---

[42] See the Guidelines for Examination in the European Patent Office, as last revised in November 2014, Part F-II, 4.9.

[43] Strasbourg Convention (n 32) art 3. Germany had been a leader of the European movement for harmonization of the patent laws and, with regard to the question of industrial application, the EPC represents to a great extent an adoption of the German approach (by adopting the wording 'susceptible of industrial application') versus the Anglo-American utility approach, see Muller and Wegner (n 9).

[44] Article 1(3) of the Paris Convention states that industrial property shall be understood in the broadest sense and shall apply not only to industry and commerce proper, but likewise to agricultural and extractive industries and to all manufactured or natural products, for example, wines, grain, tobacco leaf, fruit, cattle, minerals, mineral waters, beer, flowers and flour, see Committee of Experts, *Report of the Committee of Experts to the Committee of Ministers on the*

Preparatory work for the creation of a European Patent Convention started in 1961 with the deliberations of the European Economic Community (EEC) Patents Working Party (the Working Party).[45] The first discussions on inherent patentability were held in Brussels in 1961 and 1964, and focused on whether the proposed text should contain an express definition of invention, or whether it should follow the Strasbourg Convention approach of merely stipulating the availability of patents for new inventions that are susceptible of industrial application and involve an inventive step.[46] The conclusion finally reached by the Working Party was that inherent patentability should remain undefined in the EPC, as it had been adopted in many national laws without a 'practical problem'.[47] It is significant in this regard that the First Working Draft of the EPC dated March 1961 already contained Article 52(1) EPC in its present form.[48]

Continuing the work of the Working Party, a second period of deliberations with a higher level of participation took place at the Inter-Governmental Conference for the Setting Up of a European System for the Grant of Patents (Luxembourg Conference) between 1969 and 1972, which culminated at the Munich Conference in 1973.[49] With regard to the question of inherent patentability, the deliberations of the Luxembourg and Munich conferences focused on the types of subject matter that should be excluded from patentability,[50] rather than on whether inherent patentability should be defined further than by providing a set of patentability requirements and exclusions. The result was Article 52 of the EPC, which does not provide an express definition of inherent patentability. Instead, it reaffirms the patentability requirements set out in the Strasbourg Convention and adds an open list, more refined than the exclusions in Article 2 of the Strasbourg Convention, of subject matters that are to be excluded from patent protection.

By reproducing the same wording that the Strasbourg Convention had adopted, Article 52(1) of the 1973 EPC text provided that European

---

Meeting held at Strasbourg from 10th to 13th July of 1962, EXP/Brev (61) 8 (13 December 1961); Wadlow (n 7) 370–371.

[45] Pila (n 34).

[46] Ibid.

[47] Ibid. See also Proceedings of the 1st meeting of the PWP held at Brussels from 17 to 28 April 1961, Council of Europe Doc IV/2767/61-E (1961) Section 5, 4.

[48] Haedicke and Timmann (n 10) 91.

[49] Pila (n 34).

[50] Ibid.

patents should be granted for any inventions which were susceptible of industrial application, which were new and which involved an inventive step. In order to incorporate Article 27 of the TRIPS Agreement, this provision was amended in 2000 to specify the availability of European patents for inventions 'in all fields of technology'.[51] Nonetheless, the new wording did not change the general content and purpose of Article 52(1), or with regard to industrial application. Today, Article 52(1) of the EPC, and thus the national laws of the EPO Member States, provides that European patents shall be granted for any inventions, in all fields of technology, provided that they are new, involve an inventive step and are susceptible of industrial application. As regards the requirement of industrial application, Article 57 of the EPC, which mirrors the wording of Article 3 of the Strasbourg Convention, specifies that an invention shall be considered as susceptible of industrial application if it can be made or used in any kind of industry, including agriculture.

3.2.3 Interpretative Basis

Pursuant to Article 23(3) of the EPC, the rules that govern the interpretation of the EPC and that are binding on the EPO are the text of the EPC and the Implementing Regulations.[52] Principles of precedent are not applicable within the EPO. The EPC text and its Implementing Regulations do not refer to any prior legislation, which suggests that history is not part of their interpretative framework. Therefore, it could be concluded that the EPC, including the articles on the industrial application requirement, should be interpreted only according to the EPO's own provisions, rules and interpretative criteria. However, as an international treaty that has been part of the process of European legal harmonization, the EPC text should be subjected to other international and European agreements that can serve as interpretation guidelines.

Following this same logic, the EPO Enlarged Board of Appeal (EBA) stated in a 1984 decision that, as an international treaty, the EPC has to be interpreted in accordance with the rules of interpretation developed in the so-called 'law of nations' or public international law, including

[51] Act revising the Convention on the Grant of European Patents (European Patent Convention) of 5 October 1973, last revised on 17 December 1991, of 29 November 2000.

[52] See EPC (n 1) art 164(1).

Articles 31 and 32 of the Vienna Convention.[53] From those Articles the need is derived to interpret treaties in light of their wording and context but also their object and purpose, and if needed, preparatory work and the circumstances of its conclusion. Within the context of the EPC, this would mean that the previous works on harmonizing European patent law and the travaux préparatoires of the EPC should have legal relevance as interpretive aids.

Another relevant agreement for the interpretation of the EPC is the Paris Convention of 1990. The EPC constitutes a special agreement within the meaning of Article 19 of the Paris Convention, which contains minimum standards for substantive intellectual property law harmonization, such as, for example, rules on the right of priority.[54] In this regard, Article 87 of the EPC recognizes that a first filing made with the EPO gives rise to a right of priority under conditions and with effects equivalent to those laid down in the Paris Convention. Thus, the Convention is relevant to the EPC as for the interpretation of provisions on priority rights. On the other hand, the Paris Convention in Article 1(3) mandates the granting of patents across the whole range of industrial property. Moreover, the expression 'capable of industrial application' in the EPC is expressly and intentionally derived from the corresponding usage of the word 'industry' in the Paris Convention.[55] Therefore, although the Paris Convention does not contain substantive provisions, its conception of patent rights is relevant since its text forms an important part of the contextual background of the EPC.

Furthermore, the EPO has also confirmed that although the TRIPS Agreement could not be directly applied to the EPC, it is appropriate to take its text into consideration, since it is aimed at setting common standards regarding the availability, scope and use of intellectual property rights and gives a clear indication of more recent trends in patent law.[56] As an example, Article 27 of the TRIPS Agreement, which ensures that

---

[53] Vienna Convention on the Law of Treaties of 23 May 1969, arts 31 and 32; *Second Medical Indication/EISAI* (G 0005/83) [1984] (EPO (EBA)) points 1 to 6.

[54] See also the EPO decision in *Priorities from India/ASTRAZENECA* (G 0002/02) [2004] (EPO (EBA)) referring to the applicability of the Paris Convention rules on priority according to Article 2(1) of the TRIPS Agreement, which imposes a general obligation on WTO members to comply with the substantive provisions of the Paris Convention, regardless of whether or not a WTO member is also a party to that Convention.

[55] Wadlow (n 7) 367.

[56] *Computer program product/IBM* (T 1173/97) [1998] (EPO (TBA)) point 2.3.

different technological fields receive a treatment that is consistent with the aims of the system in each particular industry, may serve as a reference for interpreting certain provisions of the EPC. Indeed, as mentioned above, Article 52 of the EPC was amended to incorporate the mandate of Article 27 of the TRIPS Agreement, so the latter should be relevant for the interpretation of the corresponding EPC provision.

Finally, there is the Biotech Directive of 1998. Some of the provisions of the Biotech Directive have been deliberately incorporated in the EPC Implementing Regulations. In particular, the EPO has adopted the Biotech Directive's approach towards the industrial applicability of human genetic inventions as well as the Biotech Directive's exclusions of several biotech inventions on the basis of ethical concerns.[57] As Rule 26(1) (previously Rule 23b) of the EPC confirms, the Biotech Directive shall serve as a supplementary means for interpreting the Biotech Directive's provisions incorporated into the text of the EPC.

In view of the above, it seems that even though the EPC does not in theory rely on historical and external concepts as interpretative means, the EPO has recognized on several occasions the relevance of other previous and also subsequent agreements as interpretative aids. In particular, the recognition of the Vienna Convention, the Paris Convention, the TRIPS Agreement and the incorporation of the Biotech Directive are especially relevant for the interpretative framework of the requirement of industrial application. For instance, Articles 31 and 32 of the Vienna Convention confirm that the provisions of the EPC shall be interpreted teleologically. With regard to the requirement of industrial applicability, this means that Article 57 of the EPC and related provisions should be interpreted in light of their historical meaning and purpose. Further, Article 1(3) of the Paris Convention mandating that industrial property should be interpreted in its broadest sense sets a guiding standard that the EPO should follow with regard to the interpretation of the word 'industry' in Article 57 of the EPC. Article 27 of the TRIPS Agreement would be consistent with changes in the interpretation of Article 57 in response to concerns of the patent system with regard to specific technologies such as those relating to human genetic material. Finally, the incorporation of Article 5(3) of the Biotech Directive into Rule 29(3) of the EPC, which establishes the additional requirement for inventions concerning isolated human gene sequences to disclose their industrial applicability in the patent application, entails the inclusion of

---

[57] See Biotech Directive, arts 5(3) and 6 (incorporated into EPC rules 28 and 29).

the Biotech Directive as an interpretative guideline at least with regard to the EPC provisions reproducing the Biotech Directive.

3.2.4 The US Perspective

The European requirement of industrial application essentially corresponds with the US patent law requirement of utility, both concepts sharing important characteristics.[58] A comprehensive comparative analysis of the requirements of industrial application and utility falls outside the scope of this book. Nonetheless, it is significant that like the criterion of industrial application, the utility requirement has historically taken a central role in the development of the current US patent system, which further confirms the importance of this requirement in patent law.

The requirement of utility has its basis in the US Constitution, thus being a constitutional requirement and not a statutory one. 'To promote the progress of science and useful arts', Article 1 (Section 8) of the US Constitution specifically grants the Congress the power to give inventors exclusive rights to their discoveries for a limited period of time.[59] According to the Supreme Court, the basic quid pro quo contemplated by the Constitution and the Congress for granting a patent monopoly is the benefit derived by the public from an invention with substantial utility.[60] Thus, inherent in Section 8 lies the patent law bargain that in exchange for disclosing useful innovations for the benefit of the public, inventors are granted exclusive rights over their patentable inventions.

Of all the US requirements for patentability, only the utility requirement finds explicit mandate in the constitutional text. Since the passage of the first Patent Act in 1790, an inventor has been required to demonstrate that his invention is useful in order to secure a patent on it.[61] Patentable subject matter is constrained by the constitutional phrase 'useful arts'. Thus, the courts treat the utility requirement as a hybrid subject matter limitation by restricting the patent system to the useful

---

[58] See WIPO Standing Committee on the Law of Patents, *'Industrial Applicability' and 'Utility' Requirements: Commonalities and Differences* (Ninth Session, Geneva 12–16 May 2003) 14–15.

[59] US Constitution of 1787, art 1, sec 8 Powers of the Congress.

[60] *Brenner v Manson*, 383 US 519 (1996) 534–535.

[61] Donald L Zuhn, 'DNA Patentability: Shutting the Door to the Utility Requirement' (2000) 34 J Marshall L Rev 973. See also J Timothy Meigs, 'Biotechnology Patent Prosecution in View of PTO's Utility Examination Guidelines' (2001) 83 JPTOS 451.

arts.[62] In other words, they use the utility requirement to distinguish applied technology from abstract knowledge.

Apart from the Constitution, the modern US requirement of utility derives from two other sources, Congressional legislation implementing the Constitutional mandate and federal court decisions interpreting the meaning of the word 'useful' in the Constitution and in the implementing legislation.[63] In this respect, the inclusion of the utility requirement in the very wording of the Constitution and its continued vitality throughout two centuries of patent statute amendments evidence its importance and argue strongly against the possibility that the legislature intended this requirement to be easily circumvented.[64]

The concept of 'utility' itself has maintained a central place in all US patent legislation, culminating in the present law's provision.[65] The US Patents Act, in addition to setting forth the categories of patent-eligible subject matter, requires that an invention must be useful in order to receive patent protection.[66] However, the Patents Act does not provide a definition for the term 'useful' so inventors have had to rely instead on the interpretations provided by the federal courts, primarily by the Court of Appeals for the Federal Circuit (CAFC), its predecessor, the Court of Customs and Patent Appeals (CCPA) and the US Supreme Court.

The requirement of utility has not usually been an obstacle for mechanical and electrical applications, which generally have an end result or final use, but it raises greater concerns within the context of chemical and biological inventions that very often have an evolving utility, for example a general usefulness in basic research.[67] In this regard, the United States Patent and Trademark Office (USPTO) and US courts have issued conflicting decisions over how the utility test is to be applied to a chemical or biological process that produces an already known product whose utility has not yet been determined.[68] For example

---

[62] Rebecca Eisenberg, 'Analyze This: A Law and Economics Agenda for the Patent System' (2000) 53 Vand L Rev 2081.

[63] Zuhn (n 61).

[64] See Karen F Lech, 'Human Genes without Functions: Biotechnology Tests the Patent Utility Standard' (1993) 27 Suffolk UL Rev 1631.

[65] Thambisetty (n 18).

[66] US Patents Act, 35 USC, sec 101.

[67] Donald S Chisum and others (eds), *Principles of Patent Law: Cases and Materials* (3rd edn, Foundation Press 2004) 735. See also Meigs (n 61).

[68] Chisum and others (n 67) 737.

in Application of Nelson,[69] the CCPA reversed the USPTO rejection of the invention for lack of utility of a process useful for doing research on steroids, neither of which were known to have any useful results. Then in *Brenner v Manson*, the CCPA held against the USPTO decision that a process for producing certain known steroids does not need to show utility for the product as long as it is not detrimental to the public interest.[70] These decisions also contrast with the earlier more liberal notion of utility as moral or beneficial utility in 'contradistinction to mischievous or immoral', which derived from Justice Story's definition of a useful invention as one that 'should not be frivolous or injurious to the well-being, good policy, or sound morals of society'.[71]

The different interpretations of the utility doctrine were eventually discussed by the US Supreme Court in the 1966 decision in *Brenner v Manson*, which established the current utility standard by requiring patentable inventions to have substantial utility, in the sense that they provide a specific benefit in their currently available form,[72] not only as an object of scientific inquiry.[73] *Brenner v Manson* is the last case where the US Supreme Court gave a judgement regarding the question of utility. Subsequent lower court decisions have taken a more or less liberal approach towards utility. For example, in *Re Brana* the Federal Circuit reversed a rejection of claims to novel compounds that were structurally similar to a family of compounds displaying antitumour properties.[74] While in *Nelson v Bowler*, identifying a pharmacological activity of the claimed compound was sufficient for establishing a practical utility of 'real-world' value.[75] Later, in *Re Fisher* the Federal Circuit held that 'utilities that require or constitute carrying out further research to identify or reasonably confirm a "real world" context of use are not substantial utilities'.[76] Nonetheless, the leading approach (at least with regard to chemical and biological inventions) is the one established in *Brenner v*

[69] *Application of Nelson*, 280 F2d 172 (CCPA 1960) 180–181. Although a few years earlier in *Application of Bremner*, 182 F2d 216 (CCPA 1950) 217, both the USPTO and the CCPA rejected a patent application for lacking an assertion of utility and an indication of the use or uses intended.

[70] *Brenner v Manson* (n 60) 522–523.

[71] *Lowell v Lewis*, 15 Fed Cas 1018 (No 8568) (CCD Mass 1817) 1019.

[72] *Brenner v Manson* (n 60) 534–535.

[73] Ibid 529.

[74] *Re Brana*, 51 F3d 1560 (Fed Cir 1995).

[75] *Nelson v Bowler*, 206 USPQ 881 (CCPA 1980) 883.

[76] *Re Fisher*, 421 F3d 1365 (Fed Cir 2005) 1368.

Manson, which set forth the standard that has been followed by subsequent decisions,[77] and that eventually formed the basis of the current USPTO Utility Examination Guidelines.

In 1995, the USPTO established a first set of examination guidelines in response to criticism by the patent bar that patent examiners were inconsistently and inappropriately rejecting inventions involving biotechnology and therapeutic method inventions for lack of utility.[78] According to the 1995 guidelines, an asserted utility had to be specific and credible; however, nothing was mentioned about the long-established requirement of substantial utility set forth in *Brenner v Manson* and reaffirmed in subsequent decisions. In this regard, the public comments received on the Interim Written Description Guidelines on the patentability of expressed sequence tags (ESTs) suggested that the 1995 guidelines would lead patent examiners to the grant of patents based on nonspecific and nonsubstantial utilities, and that there was a need for revision or clarification of the 1995 Utility Examination Guidelines.[79] As a result, a new version of the utility guidelines was published in December 1999 along with requests for additional public comments.

In line with *Brenner v Manson* and subsequent case law, the Revised Utility Examination Guidelines added a third element to the utility test, the requirement of having a 'substantial' utility. Then following public consultation, the USPTO further amended the guidelines in 2000, although not significantly, and published a set of training materials dealing with specific technology issues, for example in the field of biotechnology. The final version of the Utility Examination Guidelines entered into force in January 2001.

In sum, the historical evolution of the criteria of industrial application and utility shows that this requirement has played a principal role in the development of patent law and the establishment of a utilitarian patent system. Nevertheless, differences exist between the roots of and approaches towards the interpretation of both concepts.[80] The next sections of this chapter will focus on the European requirement of industrial application and will include occasional references to the US criterion of utility when they are relevant for understanding certain issues and/or European decisions regarding industrial applicability.

---

[77] For a summary of other lower court decisions after *Brenner v Manson* see Meigs (n 61).

[78] Meigs (n 61).

[79] Ibid.

[80] For a further discussion about utility in international patent law see Wadlow (n 7) 355–382.

3.3 INDUSTRIAL APPLICATION: SIGNIFICANCE AND INTERPRETATION

As Lord Hoffman's decision in Biogen noted, the four conditions set out in Article 52(1), namely novelty, inventive step, capacity for industrial application and to not fall within excluded categories probably contain every element of the concept of an invention.[81] Among those four conditions, the three patent law requirements are globally recognized as the principal standards that every invention has to meet to be awarded the monopoly that a patent grants. Hence industrial application is one of the patent law requirements sine qua non an invention cannot be patented.

3.3.1 Industrial Application as a Public Policy Tool

The principal assumption underlying the creation of patent systems is that society will benefit from new technology if inventors have the incentive and reward of a patent to induce their effort and their investment in carrying out research. Thus, the ultimate purpose of patents is that in exchange for the concession of a patent monopoly, the invention needs to provide some valuable benefit to society. In this sense, the requirements (novelty, inventive step and industrial application) and conditions for granting patents reflect the terms of the deal between the inventor and society.

One of the principal objectives of patents is facilitating the practical use of inventions, including the development, application, commercialization and transfer of the patented technology. This idea flows directly from the broad instrumental or utilitarian conception of the patent system, according to which the patent system is successful to the extent that it results in obtaining more useful things for society.[82] The Strasbourg Convention's objective of promoting technical progress is also consistent with this utilitarian view of patents.[83] From this theoretical perspective, an invention would only be worthy of being patented if it

---

[81] *Biogen Inc v Medeva Plc* [1996] UKHL 18, [1997] RPC 1, point 44.

[82] See Michael R Taylor and Jerry Cayford, 'The U.S. Patent System and Developing Country Access to Biotechnology: Does the Balance Need Adjusting?' [2002] Resources for the Future 2; Dan L Burk and Mark A Lemley, 'Policy Levers in Patent Law' (2003) 89 Virginia Law Review, 1575; F Scott Kieff, 'On the Economics of Patent Law and Policy' in Toshiko Takenaka (ed), *Patent Law and Theory* (Edward Elgar 2008) 3; Robert P Merges, *Justifying Intellectual Property* (Harvard University Press 2011) 94.

[83] Strasbourg Convention (n 32) Preamble.

makes a useful contribution to society. In this sense, for an invention to be of practical use, the public needs to know how they can make and use the patented technology and what the invention is for. Therefore the criterion of industrial application, by requiring inventions to show a practical use in industry, plays a central role in meeting the patent system's utilitarian objectives as well as the interests of society.

Innovation is a public good and patent law in general is ultimately aimed at having some kind of social utility. However, not all patent law provisions are equally able to achieve such objectives. In this regard, industrial application is considered to be an essential requirement without which the claims of the patent system to promote technical and economic progress would seem to be null and void.[84] Without requiring inventions to be capable of being used by the public, society would receive contributions in the form of monopolies that cannot provide any beneficial application in practice.

Even though the reasoning behind the industrial applicability requirement appears to be rather simple, there is a certain lack of clarity as to its practical implementation. The fact that the industrial application doctrine has usually posed a low hurdle for patent applicants to clear,[85] which has resulted in an equally low number of decisions on the industrial applicability requirement, has impeded the development of a substantial body of case law on this area. It has only been with the advent of complex and research-based technologies such as genetic engineering that new rules and interpretations regarding industrial application have been developed. Consequently, there is not a long-standing test or standard that can help authorities to determine the usefulness of an invention in patent law terms. That has made industrial applicability a malleable requirement that could be used as a versatile tool for accomplishing public policy goals.

In innovation systems based on patent protection mechanisms, patent law is the principal tool that governments can use for balancing the general interests of society against the economic interests of the inventor,

---

[84] Wadlow (n 7) 358.

[85] Margaret Llewelyn, 'Industrial Applicability/Utility and Genetic Engineering: Current Practices in Europe and the United States' (1994) 16 EIPR 473; Nathan Machin, 'Prospective Utility: A New Interpretation of the Utility Requirement of Section 101 of the Patent Act' (1999) 87 California Law Review 421.

by encouraging innovation, promoting the development of new technologies and contributing to the growth of human knowledge.[86] Different preferences in the choice of patent law have arisen from political, cultural and other differences between states.[87] And many choices in patent law inherently involve policy decisions.[88] For example, depending on the level of inventive activity, authorities have traditionally moved from stricter approaches towards the application of patent law provisions to more flexible ones and vice versa. In this sense, among the different requirements and provisions stipulated by patent law, the industrial application doctrine has been remarkably adjustable and its interpretation and understanding have significantly fluctuated over time. In particular, the multiple ways of addressing the industrial applicability criterion frequently represent policy shifts to promote specific technologies.[89] For instance, a strict interpretation of the meaning of industrial application would help reduce the level of patenting activity in knowledge-based areas of technology, while a more flexible approach would allow the patentability of early research results even if their practical applications have not been fully specified.

In this regard, the requirement of industrial application can serve as a barrier to the patentability of new inventions before any practical use has been found.[90] Depending on the established policy objectives, authorities can widen or narrow the filter to allow or impede the patenting of certain claims. In other words, industrial application could be seen as a timing device that helps identify when an invention is ready for patent protection and when it is necessary to carry out further research to determine its practical value.[91] It avoids the grant of patents over inventions that may advance the industrial arts but not sufficiently to deserve the grant of the exclusive protection that a patent monopoly confers.[92] Importantly, this gating function serves to ensure that society can make use of the patents that are finally issued.

---

[86] Burk and Lemley, 'Policy Levers in Patent Law' (n 82); Sean C Pippen, 'Dollars and Lives: Finding Balance in the Patent Gene Utility Doctrine' (2006) 12 BUJ Sci & Tech L 193.

[87] Stack (n 25) 23.

[88] Ibid 23–25, 65–66.

[89] Thambisetty (n 18).

[90] Stephen B Maebius, 'Novel DNA Sequences and the Utility Requirement: The Human Genome Initiative' (1992) 74 JPTOS 651.

[91] Eisenberg, 'Analyze This: A Law and Economics Agenda for the Patent System' (n 62).

[92] Trevor Cook, *Pharmaceuticals, Biotechnology and the Law* (2nd edn, Lexis Nexis 2009) 150.

3.3.2 Article 57 EPC: Wording and Interpretation

In the European patent system, the main legal provisions addressing the requirement of industrial application (that are applicable to all fields of technology) are Article 57 of the EPC, Rule 42(1)(f) of the Regulations and Part G.III.1 of the EPO Guidelines for Substantive Examination. Article 57 of the EPC sets forth the industrial application requirement in general and the Regulations and Guidelines then further clarify, together with the EPO case law, some aspects of such provision. In this regard, the EPC Guidelines comments on Article 57 of the EPC note that the requirement of industrial application should be interpreted in very broad terms and thus, it shall only exclude very few inventions from patentability. For example, inventions that are clearly contrary to well-established physical laws. By contrast, Rule 42(1)(f) remarks on the importance of disclosing how an invention can be applicable in industry by providing that the invention's description should indicate explicitly the way in which the invention is capable of exploitation in industry only if it is not obvious from the description or from the nature of the invention. In these cases, applicants have to indicate some way in which the claimed invention can be industrially applicable. However, this requirement is only applicable when the possibility of industrial application is not clear to a person skilled in the art from the nature of the claimed invention.

The wording of Article 57 of the EPC reveals three main concepts that form the core of the patent law requirement of industrial application. This article reads that an invention shall be considered as susceptible of industrial application if it can be made or used in any kind of industry, including agriculture.

(1) The article employs the term 'susceptible', which suggests future (potential) rather than present applicability.
(2) The reference to 'industry' requires determining what fields are included within the scope of such term.
(3) Defining what kind of 'application' an invention needs to show is essential for understanding the tenor of the requirement.

3.3.2.1 Susceptible (capable)

According to the EPO Guidelines for Examination, the expression 'capable of exploitation in industry' means the same as 'susceptible of industrial application'. The difference between the words 'capable' and 'susceptible' would just be a matter of semantics and both expressions would thus be interchangeable.[93] As a synonym of the word 'capable', 'susceptible' means having the ability, fitness or quality necessary to do or achieve a specified thing, or being able to achieve efficiently whatever one has to do.[94] Thus, in contrast with the US requirement of utility, industrial application does not imply a need to show actual use (past or present).[95] In Europe, an invention that is considered to be industrially applicable is an invention that is capable or susceptible of being used by or in industry. In this respect, the reference to the term susceptible indicates the need to show future or potential use. Thus the focus of the requirement is on the idea of useful purpose, rather than actual usefulness.

Decisions like Chiron Corporation in the UK,[96] noting that the requirement of industrial application requires that the invention should relate to something that has a useful purpose, confirm that the criterion of industrial applicability is one of potential or future usefulness. However, claims with overly broad potential applications are not acceptable.[97] This question was further clarified within the context of gene patents after the incorporation of the Biotech Directive into the EPC text in 1999. The EPO decisions in Max Planck and Zymogenetics explained, in part by linking this question to issues of disclosure, that vague and speculative objectives that might or might not be achievable are not acceptable.[98] There has to be at least a prospect of a real as opposed to a purely theoretical possibility of exploitation, if it was not already obvious from the nature of the invention or from the background art. It should not be left to the skilled reader to find out how to exploit the invention by carrying out a research programme.[99] A concrete benefit or advantage must be identified, thus avoiding leaving the whole burden to the reader

---

[93] EPO Examination Guidelines (n 42) Part F-II, 4.9.

[94] Meaning of the word 'capable' according to the Oxford English Dictionary.

[95] Llewelyn (n 85); Rob J Aerts, 'The Industrial Applicability and Utility Requirements for the Patenting of Genomic Inventions: A Comparison between European and US Law' (2004) 26 EIPR 349.

[96] *Chiron Corporation v Murex Diagnostics Ltd* [1996] RPC 535.

[97] See for example *Contraceptive method/BRITISH TECHNOLOGY GROUP* (T 0074/93) [1994] (EPO (TBA)), where unsubstantiated allegations of future industrial application without any specific indication were held insufficient for fulfilling the requirement set out in Article 57 of the EPC.

[98] *BDP1 Phosphatase/MAX PLANCK* (T 0870/04) [2005] (EPO (TBA)) point 2.

[99] *Haematopoietic cytokine receptor/ZYMOGENETICS* (T 0898/05) [2006] (EPO (TBA)) point 6.

to guess or find a way to exploit an invention in industry without any confidence that any practical application exists.[100] That is, applicants must provide a clear indication of a real and achievable function.

3.3.2.2 Made or used

Article 57 requires that the invention 'can be made or used' in any kind of industry. In spite of using the conjunction 'or', this provision should not be interpreted as providing alternatives, either capable of being made or capable of being used, that enable the invention to comply with Article 57; it requires that an invention is capable of being carried out in real practice.[101] This question has often received little attention from patent authorities. It has only been with the emergence of genetic inventions, and the subsequent incorporation of the Biotech Directive in the EPC text, that more comprehensive decisions clarifying how an invention must be capable of being made or used have been developed. Before that, decisions on Article 57 were mainly focused on the concept of industry or the interface between the disclosure requirements and industrial applicability.

According to the 2004 EPO decision in Max Planck, the expression 'made or used' implies that there must be some profitable use for which the substance can be employed, in the sense that at least one industry may benefit from the invention.[102] In this regard, the decision in Zymogenetics further stressed the need to describe the subject invention in sufficiently meaningful technical terms so that it can be expected that the exclusive rights resulting from the grant of a patent will lead to some financial or other commercial benefit.[103]

However, regional and national laws and practices concerning the extent to which the practical use of the invention must be useful or profitable vary significantly. At one end of the spectrum, the requirement of industrial applicability is met as long as the claimed invention can be made in industry without taking into account the use (or usefulness) of the invention.[104] At the other end of the spectrum, in other jurisdictions the usefulness of the claimed invention is fully taken into consideration

---

[100] *Max Plank* (n 98) point 19; *Zymogenetics* (n 99) point 6.

[101] Florian Leverve, 'Patents on Genes, Usefulness, and the Requirement of Industrial Application' [2009] JIPLP 289.

[102] *Max Plank* (n 98) point 4.

[103] *Zymogenetics* (n 99) point 4.

[104] WIPO Standing Committee on the Law of Patents, *'Industrial Applicability' and 'Utility' Requirements: Commonalities and Differences* (n 58) 7.

for the determination of industrial applicability.[105] In this regard, two broadly different types of usefulness can be defined, general and specific. General usefulness would refer to the case where an invention addresses a human problem or meets a need; in contrast, specific usefulness would refer to cases when inventions actually work and produce a technical result that can be specified.[106] However, to a greater or lesser extent, the problem centres on the degree of specific usefulness that an invention shows. Hence those jurisdictions that consider all inventions that can be made industrially applicable, independently of how useful they are, would apply a clearly less demanding standard for testing the requirement of industrial application.

Nevertheless, certain common features can be identified between national practices, as well as consistency between the EPO approach and the core concepts extracted from national practices.[107] First, an invention must be applicable in any non-mental activity that belongs to the useful or practical arts that are included in the meaning of the word 'industry'. Second, the invention shall be capable of being applied in industry in the sense that it must have a useful or practical application.[108] It is not sufficient that the claimed invention can be simply made or used but an invention should have practical or useful purposes and should produce a real result to whatever extent.

Finally, the requirement that the invention can be made or used in industry also covers cases where no vendible products result. Traditionally, if an invention had any market value, it was presumed to be useful.[109] Therefore, examiners paid little attention to the requirement of industrial application because if the public could not find a profitable use for the patented invention, this one would fall into disuse;[110] and thus the patent monopoly would not impose any real restrictions in the marketplace. However, with the advances in research-based industries like gene technologies there are an increasing number of inventions that do not end in a final vendible good. In these cases, parameters other than the marketability of the product must be examined to assess the industrial application of these inventions.

---

[105] Ibid 8.

[106] Machin (n 85).

[107] WIPO Standing Committee on the Law of Patents, *'Industrial Applicability' and 'Utility' Requirements: Commonalities and Differences* (n 58) 8.

[108] Ibid 9.

[109] Joshua C Benson, 'Resuscitating the Patent Utility Requirement, Again: A Return to *Brenner v. Manson*' (2002) 36 UC Davis Law Review 267.

[110] Ibid (n 109).

In this regard, a particular question that arises when discussing what types of uses are capable of meeting the requirement of industrial application is whether the use of an invention for purely research purposes is acceptable. In most industrial fields research issues are not regarded as a problem in Europe as it is considered to be a potential form of industrial application.[111] It is only in fields like genetics that this issue becomes problematic due to the massive amount of inventions that are only useful in the context of research. In these cases, the patentability of such inventions would depend on the level of experimental evidence and specification of the function or functions of the substance concerned (see section 4.4).

3.3.2.3 Any kind of industry

The comparative study that was undertaken before the adoption of the Strasbourg Convention emphasized that the word 'industry' was generally understood in domestic laws as a defined field of economic activity that may be taken in a narrow sense, in opposition to agriculture, or in a wide sense, as a human activity oriented towards practical ends. However, even though there were differences between national practices, most countries generally admitted that inventions that were agricultural in their destination although industrial by their method of operation would fall under the scope of the law.[112] Major divergences only appeared with regard to purely agricultural activities with no industrial implications at all like breeding or harvesting.

Article 57 of the EPC provides a very flexible approach in determining the range of industries that are relevant to its norm. For the purposes of the requirement of industrial application, the notion of industry is construed in the widest possible sense and admits inventions in any kind of industry, including agriculture.[113] In this respect, the EPO Guidelines for Examination emphasises that industry should be understood in its broad sense as including any physical activity of technical character, such as activities which belong to the useful or practical arts, including enterprises in the cosmetic field, and that it does not necessarily imply

[111] Llewelyn (n 85).

[112] Committee of Experts, *Comparative Study of Substantive Law in force in the Countries Represented on the Committee of Experts on patents presented by the Secretariat-General* (n 35) 3–7, see Pila (n 2) 138.

[113] In *Schweine I/WELLCOME* (T 0116/85) [1987] (EPO (TBA)) it was held that agriculture is a kind of industry and agricultural methods are therefore, in general, methods that are susceptible of industrial application.

the use of a machine or the manufacture of an article.[114] In the same line, the case law of the EPO indicates that the term industry has to be interpreted broadly so as to include all manufacturing, extracting and processing activities of enterprises that are carried out continuously, independently and for commercial gain.[115] However, this definition of industry would not include private and personal activities. The fact that an activity is connected with professional activities does not give such act, which essentially belongs to the private and personal sphere, an industrial character.[116] Nonetheless, the range of activities that would fall outside the definition of industry is very limited in practice.

A 2003 report of the WIPO Standing Committee on the Law of Patents on the European requirement of industrial application further confirms this approach. According to the report, one common aspect to all countries is that industry is interpreted in the broadest possible sense.[117] The report also presents some examples of what countries view as industry. For instance, Sweden indicates a list of activities like commerce, forestry, public administration, gardening, hunting, fishery and defence.[118] Another example is the UK Manual of Patent Practice, which

---

[114] EPO Examination Guidelines (n 42) Part G-III, 1.

[115] See also point 5 of the Reasons in *Appetite suppressant/DU PONT* (T 0144/83) [1986] (EPO (TBA)) where the board held that enterprises in the cosmetic field – such as cosmetic salons and beauty parlours – were part of industry within the meaning of Article 57 of the EPC, since the notion of 'industry' implied that an activity was carried out continuously, independently and for financial gain. In *Thenoyl peroxide* (T 0036/83) [1985] (EPO (TBA)) the board also found that the professional use of the invention in a beauty parlour was an industrial application. See also point 2 of the Reasons in *Zymogenetics* (n 99).

[116] In *Contraceptive method/BRITISH TECHNOLOGY GROUP* (T 0074/93) [1994] (EPO (TBA)) the board found a patent application concerning the use of a contraceptive composition (a cream) for applying to the cervix of a female capable of conception not susceptible of industrial application. In contrast, in the decision for *Feminine hygiene device/ULTRAFEM INC* (T 1165/97) [2000] (EPO (TBA)) the board held that a method of using a vaginal discharge collector and disposing of the collector after a single use could be considered susceptible to industrial application if it was conceivable that these steps were carried out as a paid service and were not exclusively dependent for their execution on the instructions of the woman in question. This decision also clarified that what was relevant was the possibility that such a service might be offered by an enterprise and that the service was not one satisfying only the strictly personal needs of the woman in question.

[117] WIPO Standing Committee on the Law of Patents, *'Industrial Applicability' and 'Utility' Requirements: Commonalities and Differences* (n 58) 2–6.

[118] Ibid 4.

refers to any useful and practical activity, as distinct from intellectual or aesthetic, that is not excluded and does not restrict industry neither to tangible material nor to purely commercial or profitable activities.[119] The report leads to the conclusion that the scope of the word 'industry' in Article 57 is also interpreted by national institutions in a very broad sense so as to include almost all commercial enterprises.

3.4 FLEXIBLE VERSUS STRICT INTERPRETATIONS OF INDUSTRIAL APPLICATION

The industrial application requirement has traditionally been construed and interpreted quite broadly, so patent applicants have usually seen it as a low hurdle for patentability. Given the historical importance of the industrial application requirement, it might be expected that it would be of fully equal importance in patent law with novelty and inventiveness. However unlike the other criteria, this requirement has not been assessed against the changing state of the art, thus requiring incremental quality, but the standard of usefulness has remained broadly understood in many European countries.[120]

The EPC requirement of industrial application was not designed to impose a high barrier to patentability and as soon as some sort of practical application within a very wide range of activities is identified, industrial applicability could be acknowledged. In fact this provision has been frequently used to move aside patent applications over nonsense or very extravagant inventions. For instance, the example of a perpetual motion machine that would contravene both Newton's third law of motion and the first law of thermodynamics that is described in the EPC Guidelines,[121] or cases such as the UK Thompson's Application over an invention consisting of 'a means for purveying energy for the future comprising an engine in the form of a contra-rotational drive plate and fan that serves to create a pressure differential above and below drive plate without consuming fuel in the process'.[122]

Nonetheless, the industrial applicability criterion is an essential means for guaranteeing that society can extract some beneficial use from patented inventions, and thus a too flexible approach towards Article 57 EPC may lead to undesirable outcomes. Moreover, although patent law is

---

119 UKIPO, Manual of Patent Practice (2014) section 4, 4.02.

120 Muller and Wegner (n 9); Wadlow (n 7) 358 and 365.

121 EPO Examination Guidelines (n 42) Part G-III, 1.

122 *Thompson's Application* [2005] EWHC 3065.

technology-neutral in theory and does not distinguish between different technologies, it can be technology-specific in application.[123] The characteristics of inventions vary widely across industries and a very broad interpretation of industrial application in all cases will have difficultly meeting the utilitarian goal of the patent system in all cases. Therefore, it is important to find an adequate level of stringency in applying the industrial application standards in order to achieve the proper balance in the patent bargain.

3.4.1 Arguments in Favour of Adopting a Broad Approach towards Industrial Application

First, there is the argument that in most cases inventions will be industrially applicable because an inventor of a new device generally builds that device with a use in mind. In other words, an inventor will create a device because a specific potential use exists.[124] Thus it would be difficult to find inventions that do not have any potential use at all. It would only be in fields like chemistry or biotechnology that previously unknown random possible uses frequently arise. For example the maker of a new compound in many cases does not know the compound's ultimate use. The inventor may seek to find the cure for a disease and may discover a function that has no value for that purpose but is of great importance for a different application.[125] However in these cases it has been argued that although the final real use of the substance may not be defined until sometime after production, that substance already has useful properties from the moment of its creation.[126] Thus, the time when an industrial application is found is irrelevant as long as some useful function exists.

Second, it has been claimed that a very rigorous industrial application standard may inhibit the dissemination of information in certain fields, and that this approach would be inconsistent with those policies underlying the patent system that acknowledge the public benefits that result from the wide dissemination of information in all fields of technology.[127] The lower the industrial application standard is set, the easier it will be to

---

[123] Burk and Lemley, 'Policy Levers in Patent Law' (n 82).

[124] Eric P Mirabel, 'Practical Utility Is a Useless Concept' (1986) 36 Am UL Rev 811.

[125] Ibid.

[126] Ibid.

[127] Salim A Hasan, 'A Call for Reconsideration of the Strict Utility Standard in Chemical Patent Practice' (1994) 9 High Technology Law Journal 245.

obtain a patent, and thus the incentive to file early will be increased. Early filing promotes early disclosure and therefore, early access to patented technologies.[128] So, lower industrial application standards, by encouraging early filing, would promote early dissemination of knowledge. On the contrary, high standards would discourage or delay patent applications and thus hinder the dissemination of information.

Third, there is the argument that in a patent regime with a low industrial application requirement, duplicative research efforts would be minimized by early disclosure of the patented invention, which would encourage competitors to make arrangements with the patent owner, cease development of equal products[129] or invent around the patented product. The latter in particular is usually seen as an effective means for advancing technology.

Fourth, time and progress have dramatically altered the inventive landscape and today inventions usually offer potential benefits far earlier than commercial marketability.[130] Based on this premise, it has been argued that today inventions do not need to become final products with real practical applications in order to be industrially applicable, but that they are already useful at earlier stages. For example, discoveries that are the subject of serious scientific investigation may be sold commercially to researchers long before they have ripened into products for sale to the general public.[131] The value of inventions has changed in the last decades and new industries like biotechnology or software are responsible for the development of inventions that do not need to become a final product to be applicable but are already very useful at an early stage.

Fifth, it has been claimed that excessive industrial application standards could discourage innovation. There is little evidence that patents over inventions at an early stage of development such as research results may impede access to knowledge and future innovation.[132] However, a high standard may stifle innovation by encouraging inventors to stop carrying out research on new discoveries and keep their inventions secret if they fear that a patent will not be granted.[133] Moreover, today private funding plays a key role in research and development and a strict

---

[128] Pippen (n 86).

[129] Ibid.

[130] Benson (n 109).

[131] Rebecca S Eisenberg and Robert P Merges, 'Opinion letter as to the patentability of certain inventions associated with the identification of partial cDNA sequences' (1995) 23 AIPLA Quarterly Journal 1.

[132] Cook, *Pharmaceuticals, Biotechnology and the Law* (n 92) 150–151.

[133] Pippen (n 86).

industrial application requirement could stop investors from continuing to invest in innovative projects.

Finally, it has been argued that if an invention is not applicable in industry, it will not have commercial success and will not impose any relevant burden to society, while if the invention proves to be a commercial success, it will benefit the public. In those cases where an invention is useless to society, then the exclusive monopoly that a patent grants becomes worthless.[134] Therefore, the holder of such a patent would be taking nothing from the public because the public would not be encouraged to make, use or sell a useless invention. These patents would merely disclose information to the public;[135] therefore, there would not be any negative consequences if patents were granted on such inventions.

3.4.2 Arguments Supporting a Strict Interpretation of Industrial Application

The main argument in favour of adopting a strict approach towards industrial application is that a low standard would stifle research and development by rewarding individuals who do not contribute to innovation by providing a beneficial useful product to society. A low standard would enable researches to patent inventions without fully understanding their specific functions.[136] The holders of these patents would then be able to receive royalties from subsequent innovators who wish to carry out further research.[137] A low standard would confer patent rights to useless inventions that create a monopoly of knowledge and confer power to block off whole areas of scientific development, with no compensating benefit to the public.[138] Further, a very relaxed interpretation of industrial application would allow the grant of monopolies that would have the inevitable effect of hindering the freedom of research.[139] On the contrary, a higher standard would force inventors to continue their research until a real practical application is found. Under this view, a strict interpretation would fit better with the objectives of providing society with beneficial and useful inventions that the patent system pursues.

Second, the main policies underlying the patent system do not support a flexible standard. The incentive to invent theory is not sustainable as a

---

[134] Benson (n 109).
[135] Mirabel (n 124).
[136] Pippen (n 86).
[137] Ibid.
[138] Eisenberg and Merges (n 131).
[139] Llewelyn (n 85).

justification for a broad requirement because it does not account for today's merger of the traditionally distinct public and private spheres, which provides sufficient incentives to invent basic upstream products in many industries.[140] Likewise, neither does the incentive to disclose theory, since if a clear use has not been identified, it cannot be disclosed. A narrow industrial application requirement would strengthen patent law's incentive to disclose by allowing upstream, basic research to remain in the public domain, while promoting disclosure of downstream innovation with a real practical utility.[141] With regard to the incentive to innovate theory, which maintains that patents induce firms to carry an invention through to market, a narrow requirement enhances innovation and collaboration by allowing property rights further downstream while leaving basic discoveries in the public domain, thus promoting competitive, creative development of early research.[142] Following this argument, a loose application of Article 57 of the EPC would thus ultimately contradict fundamental policy ideas behind patent law regimes.

Third, the inventive contribution of a patent relies on its practical utility. If patents were granted over materials at the stage when they are only an experimental contribution towards a final and useful result, then they would not constitute a significant practical outcome that provides the most inventive of all the contributory steps.[143] An invention without a current or foreseeable use has no practical value, and thus patents on inventions without an identified utility would give rise to useless monopolies. Correspondingly, there is no sense in attributing the right to exclude others from making, using, selling or importing an invention that is worthless.[144] In this regard, a stringent approach towards industrial application serves to guarantee that a real practical contribution to industry exists, or in other words, a higher standard would protect society from monopolies on useless inventions.

Finally, philosophical approaches to patent law also support a strict standard. For instance, the labour-based theory supports a narrow interpretation providing that property rights should only be granted to things

---

[140] Teresa M Summers, 'The Scope of Utility in the Twenty-First Century: New Guidance for Gene-Related Patents' (2002) 91 Georgetown Law Journal 475.

[141] Ibid.

[142] Ibid.

[143] William Cornish, David Llewelyn and Tania Aplin, 'Intellectual Property in Biotechnology' in *Intellectual Property: Patents, Copyright, Trade Marks & Allied Rights* (7th edn, Sweet & Maxwell 2010) 917.

[144] Machin (n 85).

that add value to society and do not harm common resources. Within the context of patent law, Locke's notion of awarding property rights would support a narrow requirement according to which the public should retain the right to access 'upstream' products such as research tools while property rights are to be granted for 'downstream' innovations which are of value to society.[145] This approach is primarily concerned with protecting the interests of society wherever property rights exist. Therefore, ensuring that patented inventions provide a realizable benefit to society essentially concurs with this approach.

3.5 CONCLUDING REMARKS

The requirement of industrial application has historically played a key role in achieving fundamental patent policy objectives. By requiring that inventions provide a real practical contribution to the industrial arts, the industrial applicability criterion helps to ensure that society can ultimately benefit from patented technologies. However, industrial application has not always received much attention in contemporary patent legal and policy discussions. Despite its historical importance as an essential pillar in the justification of patent systems, this requirement has not generally posed a significant challenge to patent applicants. With the adoption of the EPC it was clarified that this requirement should only impose a barrier to patentability in extremely limited situations. Nonetheless, a very flexible approach towards interpreting industrial application would ultimately encourage the patentability of inventions that are incapable of performing a practical function in industry. Thus, it seems important to find an adequate level of stringency in applying the industrial application standards in order to achieve an optimal balance between the private and public interests at stake in patent systems.

The purpose of this chapter was to offer a comprehensive understanding of the history, significance and interpretation of the requirement of industrial application in the European patent system. The chapter sets the basis for analysing the changes introduced by the Biotech Directive regarding the interpretation of industrial application for inventions covering human DNA fragments.

---

[145] See Summers (n 140).

4. The industrial applicability of human genetic inventions

4.1 INTRODUCTION

Following the important developments in molecular biology that took place during the 1990s, concerns arose regarding the growing number of patent applications over human DNA sequences and related proteins of unknown function that were not capable of showing any application in industry. To deal with this issue, the Biotech Directive included several provisions requiring applicants to disclose the industrial application of inventions concerning isolated human gene sequences at the time of filing the patent application. The new standard was later incorporated into the Implementing Regulations to the EPC,[1] and as a result, a full new body of case law on the industrial application of gene sequences and encoded proteins has been developed in recent years. This chapter first studies the main scientific concepts underlying the creation of inventions based on human genetic information and then examines the patenting landscape of isolated human DNA fragments and their corresponding proteins and the demand for a response to the growing number of patent claims over human gene-related inventions of unknown use. Second, the chapter deals with the justifications for adapting patent systems to new technologies to then explore the background of Article 5(3) of the Biotech Directive and its incorporation into the EPC system. Third, the chapter examines the EPO case law on Rule 29(3) of the EPC. Finally, the chapter analyses the case of *Eli Lilly v Human Genome Sciences* (HGS) before both the UK Supreme Court and the EPO, and the implications of this case for the biotech industry.

[1] Implementing Regulations to the Convention on the Grant of European Patents of 5 October 1973.

4.2 THE PATENT CLAIM: PROTEIN-CODING GENES OF UNKNOWN FUNCTION

4.2.1 The Scientific Concepts

Since the discovery of the DNA structure and recombinant DNA techniques the principal pursuit of biotech research has been to decipher genes, entire genomes and the proteins they can produce.[2] In consequence the inventive landscape within the biotechnology field started to receive more and more inventions, and patent applications, concerning human genetic material. DNA, which can be described as the code through which living organisms pass their traits to offspring, and RNA, a related molecule made from DNA, are the two basic kinds of molecules that carry the genetic code, and that, through specific regions or sequences called genes, provide cells with the instructions for producing or coding proteins.[3] DNA is contained in the nucleus of almost every human and mammalian cell with the exception of very basic forms of life such as bacteria.

A polypeptide or protein is a chain, or (sometimes) linked chains, of amino acids that performs many essential functions in the body. Proteins are polymers, that is, molecules that contain many copies of a smaller building block, covalently linked by peptide bonds.[4] Such building blocks are α-amino acids that are specified by the genetic code, of which there are 20 that occur regularly in the proteins of living organisms.[5] A protein typically contains from 50 to 2000 amino acids, there being proteins consisting of approximately 30 000 amino acids.[6] One protein differs from another based on the sequence of amino acids, and it is this sequence of amino acids that determines the protein's ultimate function. Because of their many different shapes and chemical properties, proteins can perform a wide range of functions inside and outside the cells that either are essential for life, or provide selective evolutionary advantage to

---

2 Eric S Lander and others, 'Initial Sequencing and Analysis of the Human Genome' (2001) 409 Nature 860.

3 Ibid.

4 James D Watson (ed), *Molecular Biology of the Gene* (7th edn, Pearson 2014) 121.

5 Ibid 121.

6 Herbert Zech and Jürgen Ensthaler, 'Stoffschutz bei gentechnischen Patenten – Rechtslage nach Erlass des Biopatentgesetzes und Auswirkung auf Chemiepatente' [2006] GRUR 529.

the cell or organism that contains them,[7] or provoke disease. Thus characterizing the structures and functions of proteins is fundamental for understanding how cells work. Genes activate the production of proteins by bringing together different combinations of amino acids in specific orders. Different genes attract different amino acids that in turn combine to produce different proteins, which make cells function in different ways.[8] A gene normally includes non-coding regions (introns) as well as protein encoding regions (exons). The first step in gene expression is removing the non-coding regions of RNA as the RNA is transcribed into mature messenger RNA (mRNA), which contains the protein-encoding regions of a gene.[9] The mRNA of an organism can then be isolated and copied to a mirror image of itself in a complimentary and more stable DNA sequence (cDNA), which has the same sequence of nucleic acids as the original piece of DNA in the organism's genome.[10] When an organism is discovered to produce a protein that could be useful in other contexts, this protein is isolated and a small portion of its amino acid sequence is determined in order to identify which particular gene codes for that protein.[11] Once the gene has been identified, it can be used to artificially produce the desired protein.

---

[7] Harvey F Lodish and others (eds), *Molecular Cell Biology* (7th edn, WH Freeman and Co 2013) 59.

[8] Hector L MacQueen, *Contemporary Intellectual Property: Law and Policy* (2nd edn, Oxford University Press 2011) 510–511.

[9] Lander and others (n 2). See also David L Lockhart and Elisabeth A Winzeler, 'Genomics, Gene Expression and DNA Arrays' (2000) 405 Nature 827. The coding sequence of a protein-encoding gene is a series of three nucleotide codons that specifies the linear sequence of amino acids in its polypeptide product. In eukaryotic cells the coding sequence (exons) is periodically interrupted by stretches of non-coding sequence (introns). Once transcribed into an RNA transcript the introns must be removed and the exons joined together to create the mRNA for that gene. The process of intron removal is called RNA splicing and converts the pre-mRNA into mature mRNA, see Watson, *Molecular Biology of the Gene* (n 4) 467–468.

[10] Lander and others (n 2); J Craig Venter and others, 'The Sequence of the Human Genome' (2001) 291 Science 1304.

[11] From this information, the cDNA sequence encoding the protein can be discovered, isolated and inserted into a host through an artificial process, usually a bacteria or cell line, which then replicates and produces many copies of the cDNA sequences. These collections of cDNA sequences inserted into hosts are called cDNA libraries, which are collections of the reference material (DNA) encoding the proteins made and used by that cell, see Christopher A Michaels, 'Biotechnology and the Requirement for Utility in Patent Law' (1994) 76 JPTOS 247.

In the living organism, only parts of the mRNA are translated into a protein. It is the genetic code that specifies which amino acids are added to the protein based on the mRNA sequence.[12] Given the existence of 20 amino acids but only four bases, groups of several nucleotides must somehow specify a given amino acid.[13] However, even with only four bases, the number of potential DNA sequences is very large for even the smallest of DNA molecules; a virtually infinite number of different genetic messages can exist.[14] Moreover, some of these amino acids can modify when already part of a protein.[15] Researchers thus need to identify the sequence of nucleotides in DNA that encodes the amino acid sequence of a particular protein, which is the encoding gene of that protein. There appear to be about 30 000 to 40 000 protein-encoding genes in the human genome.[16] These genes are more complex than in other organisms, with more alternative splicing generating a larger number of protein products,[17] some genes being able to produce up to 25 000 different proteins by combinatorial splicing.[18] The coding DNA regions are therefore the carriers of most of the genetic information that is responsible for phenotypic functions. However, they represent only between 3 to 5 per cent of the total DNA.[19] ESTs, which are short subsequences of a cDNA sequence, identify only these interesting coding fragments, selecting the main portions of the human genome. EST

---

[12] Lockhart and Winzeler (n 9); Denis Schertenleib, 'The Patentability and Protection of DNA-Based Inventions in the EPO and the European Union' (2003) 25 EIPR 125. Translation of the mRNA requires RNA adaptor molecules called tRNAs. Each tRNA contains a sequence of adjacent bases (anticodon) that bind specifically during protein synthesis to successive groups of bases (codons) along the RNA template, see Watson, *Molecular Biology of the Gene* (n 4) 34.

[13] Watson, *Molecular Biology of the Gene* (n 4) 37.

[14] Ibid 30–31.

[15] Ibid 121. See also Zech and Ensthaler (n 6).

[16] Lander and others (n 2).

[17] Ibid. See also Edmund Tischer and others, 'The Human Gene for Vascular Endothelial Growth Factor. Multiple Protein Forms Are Encoded through Alternative Exon Splicing' (1991) 266 Journal of Biological Chemistry 11947; Lockhart and Winzeler (n 9). It is estimated that 90 per cent or more of the protein-encoding genes in the human genome are spliced in alternative ways to generate more than one isoform (the alternative products when there is more than one polypeptide product are called isoforms), see Watson, *Molecular Biology of the Gene* (n 4) 469.

[18] Andreas Oser, 'Patenting (partial) Gene Sequences Taking Particular Account of the EST Issue' (1999) 30 IIC 1; Schertenleib, 'The Patentability and Protection of DNA-Based Inventions in the EPO and the European Union' (n 12).

[19] Oser (n 18).

cDNAs do not usually encompass the entire sequence of the original mRNA. They are usually 200 to 500 nucleotides long, and are generated by sequencing either one or both ends of an expressed gene, and consequently do not give complete DNA sequence information. Nonetheless, experiments have traditionally included the use of ESTs in attempts to discover and map human genes,[20] and the novel proteins they code for. Then in the 1990s the traditional 'wet lab' approach to biotechnology was substituted by a new computer-assisted approach called 'bioinformatics', which allows researchers to identify and isolate gene sequences and the corresponding proteins by comparing sequence homologies with previously identified and characterized genes.[21] Use of sequence comparisons between different proteins to deduce protein function has expanded substantially in recent years as the genomes of more and more organisms have been sequenced.[22] However, notwithstanding the value of these comparisons for acquiring an insight into protein structure and function, researchers have to be careful when attributing to one protein, or a part of a protein, a function or structure similar to another based only on amino acid

---

[20] Lander and others (n 2); Venter (n 10).

[21] Miguel A Andrade and Chris Sander, 'Bioinformatics: From Genome Data to Biological Knowledge' (1997) 8 Current Opinion in Biotechnology 675; Ian M Cockburn, 'State Street Meets the Human Genome Project: Intellectual Property and Bioinformatics' in Robert W Hahn (ed), *Intellectual Property Rights in Frontier Industries: Software and Biotechnology* (AEI-Brookings Joint Center for Regulatory Studies 2005) 114; Robert Fitt and Edward Nodder, 'Specific, Substantial and Credible? A New Test for Gene Patents' (2006) 9 BIO-Science Law Review 183. This increase in the use of bioinformatics has been driven by the increase in genetic sequence information and the need to store, analyse and manipulate the data. There are now a huge number of sequences stored in genetic databases from a variety of organisms, including the human genome. This genetic information is an essential starting point for molecular biology research, with the main primary databases being GenBank at the National Institutes of Health (NIH) in the US, the European Molecular Biology Laboratory (EMBL) at the European Bioinformatics Institute (EBI) at Cambridge UK and the DNA Database of Japan (DDBJ) at Mishima in Japan. One of the most useful bioinformatics resources is Basic Local Alignment Search Tool (BLAST), which allows the submission of a DNA sequence via the internet and compares it to all the sequences contained within a DNA database in order to identify sequences of similarity, see Keith Wilson and John M Walker (eds), *Principles and Techniques of Biochemistry and Molecular Biology* (7th edn, Cambridge University Press 2009) 170 and 345.

[22] Lodish and others (n 7) 69.

sequence similarities.[23] There are examples in which proteins with similar overall structures participate in different biological processes and display different functions as well as cases in which functionally unrelated proteins with dissimilar amino acid sequences have very similar folded structures.[24] The reason for this is that whereas the number of genes is strictly limited, the functions and utilities of those genes are not, and therefore the value of ESTs is limited since they only represent a part of the original DNA sequence.

As explained above, the sequence of DNA is what specifies the sequence of amino acids in a protein. In this regard, bioinformatics help to predict the amino acid sequence of a protein from the cDNA that codes for it. That is, in order to determine the structure and properties of such protein, researchers can use computer-assisted techniques to compare sequences and, if two proteins share similar sequences across regions, then assume that they will have similar structure and properties. However, these are just suppositions. The amino acid sequence of a protein might be used to assign the protein to a particular protein family or superfamily, the members of which often but not always have a significant degree of homology and have similar functions.[25] Therefore, although this information could provide an indication of the function of the protein in question, in some cases proteins have a wide range of different functions even when they are part of the same family.[26]

Thus, the fact that a protein can be ascribed to a particular family does not necessarily provide a clear indication of its function. Protein families also evolve by duplication, mutation and the exchange of whole functional domains; for example, immunoglobulin receptors involved in immunity share sequence similarities with various cell-signalling receptors that are involved in nervous system development such as nerve growth factor.[27] Their structures may show similarities but their functions

---

[23] Wilson and Walker (n 21) 346; Zech and Ensthaler (n 6); Lodish and others (n 7) 69.

[24] Lodish and others (n 7) 69.

[25] See Lockhart and Winzeler (n 9); Lander and others (n 2); Venter and others (n 10).

[26] See Lockhart and Winzeler (n 9); Lander and others (n 2); Venter and others (n 10); Matthew Spencer, 'What's the Use?!' (2009) 10 BIO-Science Law Review 56; Andrew Sharples, 'Industrial Applicability for Genetics Patents – Divergences between the EPO and the UK' (2011) 33 EIPR 72.

[27] Schertenleib, 'The Patentability and Protection of DNA-Based Inventions in the EPO and the European Union' (n 12).

differ.[28] Therefore, it would be inaccurate to hold that by uncovering the existence of a protein its function would always be immediately known, although in some cases it might be possible to make an 'educated guess'.

4.2.2 The Patenting Landscape in the 1990s: A Growing Number of Claims over Genetic Fragments of Unknown Function

There are 2.9 billion base pairs in the human genome, which are arranged into approximately 30 000 genes. The complete map of the human genome was completed in 2003, however, today scientists still do not know the properties of many genes or what proteins they can produce, and so their functions remain uncovered.[29] Nonetheless, ascertaining the role of a genetic sequence is essential for treating multiple diseases since many genetic disorders occur when one or more nucleobases are missing or misplaced, or when genes interact with each other or the environment in adverse ways.[30] The ultimate benefit from genetic research does not lie in the disclosure of the obvious, that is, that a gene can produce a protein, but in the deliberate discovery and identification of an important function that is useful for society; for example, that chemokine receptors can function as human immunodeficiency virus (HIV) co-receptors.[31] The challenge is to match the genetic codes with their practical applications in order to better understand how genes work, what role they play in our bodies and how they can be used to cure disease.[32] If no practical use is disclosed, it is less justifiable to allow the grant of exclusive monopoly rights that may simply impose barriers to further research. Patenting gene parts without a known function would mean a departure from the traditional patent bargain that allows a patent on an invention in exchange for the disclosure of a useful contribution to the art. This is also true in other fields of technology; however, given the importance of genetic research in human healthcare and other high impact areas, awarding patents that are not readily useful would pose an unnecessary threat to the progress of research in this field and thus, for example, to the development of new cures.

---

[28] Ibid.

[29] MacQueen (n 8) 511.

[30] Ibid.

[31] Mattias Luukkonen, 'Gene Patents: How Useful Are the New Utility Requirements' (2000) 23 Thomas Jefferson Law Review 337; Li Westerlund, *Biotech Patents: Equivalence and Exclusions under European and U.S. Patent Law* (Kluwer Law International 2002) 17.

[32] MacQueen (n 8) 511.

The first patents on human DNA sequences issued in the 1980s did not raise too many concerns because they related to long-awaited drugs such as granulocyte colony stimulating factor or erythropoietin.[33] Inventions usually met the patentability criteria because researchers generally started research with a known protein with activities of interest and used that information to identify the encoding gene, which was a difficult and laborious process.[34] However, advances in the identification of the human genome generated worldwide entry of patent applications over gene-related inventions,[35] many of which did not disclose any function.[36] Today patent protection is often sought for methods and substances that relate to isolated DNA and/or RNA sequences, as well as the proteins encoded by them,[37] even if their functions have not been determined. This is because the value of gene sequences mainly resides in the proteins they can produce, which are the ones that control the functioning of the human body, and thus obtaining a patent for these substances gives the patentee the power to charge a royalty to everyone who wants to access this information.

In attempting to sequence the human genome, scientists of the HGP first sequenced short portions of human DNA sequences (ESTs) to use them as tools for sequencing entire genes.[38] The use of EST technology to locate protein-coded gene sequences was technically refined in the 1990s and brought to an efficient large scale on the basis of methods that were developed in the 1970s and 1980s.[39] However, as explained above, ESTs are small subsequences of cDNA that cannot code for a complete,

---

[33] Rob J Aerts, 'The Industrial Applicability and Utility Requirements for the Patenting of Genomic Inventions: A Comparison between European and US Law' (2004) 26 EIPR 349.

[34] Martin Enserink, 'Patent Office May Raise the Bar on Gene Claims' (2000) 287 Science 1196. See also Fitt and Nodder, 'Specific, Substantial and Credible? A New Test for Gene Patents' (n 21); Spencer (n 26).

[35] Amanda Warren-Jones, *Patenting rDNA: Human and Animal Biotechnology in the United Kingdom and Europe* (Lawtext Pub 2001) 19.

[36] Oser (n 18).

[37] Timo Minssen and David Nilsson, 'The Industrial Application Requirement for Biotech Inventions in Light of Recent EPO & UK Case Law: A Plausible Approach or a Mere "Hunting License"?' [2012] EIPR 689.

[38] Lander and others (n 2); M Scott McBride, 'Patentability of Human Genes: Our Patent System Can Address the Issues without Modification' (2001) 85 Marq L Rev 511; Venter and others (n 10).

[39] Lander and others (n 2); Venter and others (n 10). See also Oser (n 18).

functional protein or even any part of a protein.[40] The interest in them comes from the idea that they could be the subject of patent applications claiming not only the ESTs but also any gene associated with the ESTs, any protein encoded by the genes and any antibodies that bound to the proteins.[41] Thus, in the 1990s companies like Incyte, Celera Genomics, and Human Genome Sciences filed tens of thousands applications over ESTs or single base-pair mutations (single nucleotide polymorphisms (SNPs)),[42] many of which did not disclose well-defined functions like the function of related proteins or the industrial applicability of the claimed gene sequence.

In particular, the 1991 patent application by the US NIH over ESTs at the US Patent and Trade Mark Office (USPTO) represented a turning point in the discussions over the industrial applicability of human gene patents. In June 1991 the NIH filed a patent application involving partial cDNA sequences and ESTs.[43] This application was not released to the public but from the relevant literature it can be deduced that it included claims relating to the identification of approximately 6800 partial cDNA sequences, the so-called ESTs (around 340) and their protein products.[44] Both kinds of sequences were claimed to be of practical utility based upon several uses, for instance, as genetic markers for forensic identification or tissue typing.

This and other patent applications at that time suggested that scientists were using rapid screening and sequencing of large numbers of DNA sequences to randomly sequence segments of human DNA without knowing what protein they coded for.[45] Today, it is still often the case that the only known utility of an isolated and purified DNA or RNA sequence is its scientific significance as an object for further research. Their real useful functions can, if at all, only be implemented as a result of increased intellectual and expert efforts.[46] Unlike fully sequenced genes with known and exploitable functions, for example for specific

---

[40] Sven Bostyn, 'The Patentability of Genetic Information Carriers' [1999] IPQ 1; Paula K Davis and others, 'ESTs Stumble at the Utility Threshold' (2005) 23 Nat Biotech 1227.

[41] Davis and others (n 40).

[42] Cockburn (n 21) 112–113.

[43] J Craig Venter and Mark Adams, USPTO Patent Application No 07/716831, Sequences (applied 20 June 1991).

[44] Bostyn 'The Patentability of Genetic Information Carriers' (n 40).

[45] Stephen B Maebius, 'Novel DNA Sequences and the Utility Requirement: The Human Genome Initiative' (1992) 74 JPTOS 651.

[46] Oser (n 18).

therapeutic and diagnostic uses, most ESTs only serve to discover new genes, identify an expressed gene of unknown function or as a marker for localisation of a gene on a physical map of genome.[47] In consequence, patenting genes started to look less like rewarding the creation of useful downstream products and more like patenting scientific information. Although other organizations filed similar patent applications in the 1990s and maintained those filings,[48] the NIH later reversed its position on ESTs patents and abandoned its applications. A few years later, on 6 October 1998 the USPTO issued the US Patent No 5817479 for 'Human Kinase Homologs', the first patent known to include ESTs, to Incyte Pharmaceuticals, Inc.[49]

Similarly in Europe, soon after the EPO started to accept patent applications in June 1978,[50] the first attempts were made to patent important molecular biology achievements. However, determining the exact amount of applications and patent grants over human genes is somewhat difficult due to the fact that the International Patent Classification (IPC)[51] does not provide a specific mark for isolated human gene fragments. The IPC does not distinguish between human or animal genes, and patents over genetic information are generally assigned more than one mark. The most relevant classes include C12N and C07K. C12N covers inventions within genetic engineering relating to DNA and RNA, while C07K relates to proteins and peptides and may cover genetic sequences that encode proteins. More specifically, C12N15/12 to C12N15/28 describe genes encoding animal proteins.

Between 1978 and December 1997, a year before the adoption of the Biotech Directive, the EPO published 1251 patents relating to genes encoding animal (including human) proteins. Some examples are the European patent EP0411946 over DNA encoding human GP 130 protein or EP0431065 regarding a full-length human laminin binding protein cDNA sequence. A more specific search in the EP database shows that

---

[47] Aerts, 'The Industrial Applicability and Utility Requirements for the Patenting of Genomic Inventions: A Comparison between European and US Law' (n 33).

[48] Davis and others (n 40).

[49] HUGO Intellectual Property Committee Statement on Patenting of DNA Sequences in Particular Response to the European Biotechnology Directive (April 2000).

[50] The EPC was adopted in 1973 but the EPO did not start to accept patent applications until 1978.

[51] The International Patent Classification was established by the Strasbourg Agreement Concerning the IPC of 1971 (1971) <http://www.wipo.int/classifications/ipc/en/>.

within the same time period, the EPO published approximately 408 patents with the words 'DNA' or 'human' or 'gene' in the title, with both the words 'DNA' and 'human' in the title or abstract and C12N15 as the IPC classification. For its part, the Gowers Review reported in 2006 that patents had already been granted for almost 20 per cent of human gene DNA sequences by patent offices worldwide, 4382 out of the 23 688 known human genes.[52]

Because many of these inventions rarely disclosed a substantially specific practical use and also raised doubts about their novelty and inventive character, these patents became controversial because they were perceived as low-quality patents that were not able to meet the patent law criteria.[53] In particular, the question of industrial applicability has gradually become an integral part of the policies regarding the patentability of DNA-related inventions.

4.2.3 The Inventor's Perspective versus the Public Response to the Patenting of Gene Sequences of Unknown Utility

Despite being aware that many genetic inventions are not industrially applicable, in general biotech companies have continued to stress the need to protect their work at a stage when an actual use cannot be shown but where there is evidence that a significant use will follow from further research. The main argument is that although genes per se are generally of limited direct application, they are of substantial value since they enclose information that can give rise to the development of important innovations in essential sectors such as healthcare.[54] Under this view, the contribution to the art of genes goes beyond their industrial application. Thus patenting newly discovered genes of unknown functions would still encourage inventors to continue research in this field.

Another frequent claim relates to the long duration between the discovery of a specific gene sequence and the final realization of a useful product. For example ESTs, unlike penicillin, are not medications themselves but need further transformation before they can be used in practice.[55] However, companies consider patent protection to be important before they have been able to identify a specific industrial use, so

---

[52] See Andrew Gowers, *Gowers Review of Intellectual Property* (Independent report, UK Government 2006) para 2.30.

[53] Suzanne Scotchmer, *Innovation and Incentives* (MIT Press 2004) 75.

[54] Florian Leverve, 'Patents on Genes, Usefulness, and the Requirement of Industrial Application' [2009] JIPLP 289; Minssen and Nilsson (n 37).

[55] Sean C Pippen, 'Dollars and Lives: Finding Balance in the Patent Gene Utility Doctrine' (2006) 12 BUJ Sci & Tech L 193.

they can start recouping their investments and continue developing a finally useful product.

In this regard, because of the highly expensive and competitive nature of genetic research, companies try to obtain patent protection at the earliest possible stage, before the exact applications or functions of the substances are known.[56] Their argument is that if it can be established that an invention might be capable of use in the future, once further research has been undertaken, then this should prove sufficient to allow a patent grant.[57] With patent protection, companies would then reinforce their position in the market and would attract more potential investors.

By contrast, patent authorities, interest groups and other stakeholders worldwide have expressed concerns against the increasing patenting of genes or parts of genes of unknown use. In the early 1990s the USPTO began rejecting patent applications over inventions of potential therapeutic value for lack of utility if the patent applicant could offer no proof of clinical efficacy.[58] Nonetheless in February 1997, after having refused the NIH applications, the USPTO decided that ESTs were patentable even if their only utility was:

- as biochemical probes or generally as research tools[59]
- for forensic identification, tissue type or origin identification, chromosome mapping and identification and
- to tag a gene of known and useful function.[60]

However, a strict interpretation of the utility requirement has persisted in general terms in the US, and ultimately led to the implementation of new utility guidelines aimed at reinvigorating the utility requirement. In this regard, after implementing its Utility Examination Guidelines in January 2001, which reflected the reasoning of the US Supreme Court in *Brenner v Manson*,[61] the USPTO began to frequently reject ESTs

---

[56] Minssen and Nilsson (n 37).

[57] Margaret Llewelyn, 'Industrial Applicability/Utility and Genetic Engineering: Current Practices in Europe and the United States' (1994) 16 EIPR 473.

[58] Rebecca Eisenberg, 'Analyze This: A Law and Economics Agenda for the Patent System' (2000) 53 V and L Rev 2081.

[59] Davis and others (n 40).

[60] Dorothy R Auth, 'Are ESTs Patentable?' (1997) 15 Nat Biotech 911.

[61] *Brenner v Manson*, 383 US 519 (1996).

patent applications.[62] The principal question with regard to the utility of genetic material is basically whether patents should be granted in exchange for some new information for which current practical value cannot be established.

Concerns against granting patent protection to gene sequences with no known function have also been raised by biomedical researchers and patient advocates. For example Dr James Watson, co-discoverer of the DNA structure, made public his disagreement with the NIH position towards the patentability of newly discovered ESTs.[63] Further, Dr Daniel Cohen, one of the pioneers behind the HGP, said that he did not believe it was right to file such patent applications by stating that 'if you have no knowledge of the function of the sequence, then you should let people compete freely to discover it'.[64] These reactions ultimately show the importance of not overlooking the requirement of industrial application for the advance of science.

With regard to the policy discussions attempting to define the appropriate scope of DNA patent claims and the invention's utility, Lisa Feisee from the Washington, DC-based Biotechnology Industry Organization (BIO) noted the problematic character of this issue and held that: 'It's not very complicated to find ESTs. It's not like you've done invention of any sort, basically anyone can do it.'[65] According to this view, the real contribution of patents over ESTs would reside on their practical utility and not on their mere existence in isolation.

Further, the 1995 report of the National Academies Policy Advisory Group (NAPAG) of the Royal Society stated that merely isolated genes should fall under the sphere of basic discoveries and that patents should only be granted to achievements that had been carried far enough to demonstrate practical results.[66] In this regard, the authors of the NAPAG report considered that the EPC requirement of industrial application might be too broad to ensure a clear separation between the spheres of discovery and invention, and recommended the introduction of a separate

---

[62] The first appeal of such a rejection, *Re Fisher* 421 F3d 1365 (Fed Cir 2005), was argued before a three-judge panel of the United States Court of Appeals for the Federal Circuit in May 2005, see Davis and others (n 40).

[63] Llewelyn (n 57).

[64] Quoted in Llewelyn (n 57).

[65] Quoted in Stacey Lawrence, 'US Court Case to Define EST Patentability' (2005) 23 Nat Biotech 513.

[66] National Academies Policy Advisory Group (NAPAG), *Intellectual Property & the Academic Community* (The Royal Society 1995) 18–19. See also Leverve (n 54).

requirement of utility into the EPC. In a subsequent report, however, the Royal Society held that the growing tendency to file for patents over scientific knowledge with no practical information could be addressed by applying the requirements of industrial application and sufficiency more rigorously, for example by implementing the 'specific, substantial and credible' test developed in the US.[67]

A few years later, the Nuffield Council also adopted a fairly critical position regarding the patenting of genetic sequences, such as ESTs and SNPs, of unclear function.[68] Considering the deleterious effects that these patents (especially those concerning research tools) may have on further progress of genetic research and the successful exploitation of its results, the Nuffield Council expressly stated that this practice should be discouraged, recommending the review and strengthening of the requirement of industrial application so that the grant of a patent reflects more properly the inventor's contribution.

Then in 1997, the Human Genome Organisation (HUGO) published a statement expressing their concern that patenting ESTs of uncertain functions would make subsequent innovation unduly dependent of such patents.[69] In particular, the HUGO 1997 statement stressed its agreement with allowing patents when a specific function has been identified, that is, the useful benefits of genetic information, but explicitly opposed the patenting of randomly isolated gene sequences encoding proteins without specified utility, for example, 'as probes to identify specific DNA sequences'.[70]

The concerns expressed from both sides of the spectrum with respect to the industrial application of isolated human genes demanded a response addressing both public and private interests. In this case the relevant question is whether the patent practice should adhere to the conventional conceptions or rather, how the existing concepts should be applied to fit in best with the technology.[71] With regard to the issue of patenting genes of unknown function, the requirement for an invention to be industrially applicable addresses this specific challenge, and seems therefore to be the most appropriate framework within which the

[67] Royal Society Working Group on Intellectual Property, *Keeping Science Open: The Effect of Intellectual Property Policies on the Conduct of Science* (The Royal Society 2003) 12.

[68] Nuffield Council on Bioethics, *The Ethics of Patenting DNA* (Nuffield Council on Bioethics 2002) 71–72.

[69] HUGO Intellectual Property Committee Statement on Patenting Issues Related to Early Release of Raw Sequence Data (May 1997).

[70] Ibid.

[71] Westerlund (n 31) 9.

patentability of such genes should be dealt with.[72] Within this context the Biotech Directive introduced a set of specific provisions raising the bar of industrial applicability in order to ensure that inventions related to human genes are only patented if they actually show a clearly defined industrial application.

4.3 INDUSTRIAL APPLICABILITY OF GENE FRAGMENTS: THE BIOTECH DIRECTIVE'S APPROACH

4.3.1 Modifying Existing Standards to Accommodate New Technologies

There has been much debate regarding the extent to which existing patent law regimes need to be revised to accommodate new frontier technologies like biotechnology. In most countries inventions in this field are not subject to special provisions and they have been accommodated by general patent law standards. However, in Europe the disparity in approaches between Member States, for example the Netherlands' refusal to apply patent law to biological material, triggered the Commission's initiative to enact a specific directive dealing with the patentability of biotechnological inventions. The 1988 proposal for a directive was part of the wider strategic goals of completing the internal market and promoting the European biotechnology sector.

There are different alternative approaches for dealing with the various issues that arise from the patenting of genetic material, including:

- the creation of a sui generis form of protection
- imposing compulsory licensing for specific gene patents
- raising the threshold of patent law standards, noting that patent law criteria operate differently, and have different importance in different fields of the creative environment[73] or
- avoiding the patenting of genetic information at all.[74]

---

[72] Ng-Loy Wee Loon, 'Patenting of Genes – A Closer Look at the Concepts of Utility and Industrial Applicability' [2003] IIC 393; Bonnie W McLeod, 'The "Real World" Utility of miRNA Patents: Lessons Learned from Expressed Sequence Tags' (2011) 29 Nat Biotech 129.

[73] Scotchmer, *Innovation and Incentives* (n 53) 97.

[74] See John M Golden, 'Biotechnology, Technology Policy, and Patentability: Natural Products and Invention in the American System' (2001) 50 Emory LJ

In order to address concerns over the patenting of gene fragments of unknown function, the industrial application requirement was found to be the most suitable means of doing so at the time the Biotech Directive was adopted. In this regard, the Biotech Directive introduced Article 5(3) (and related recitals), imposing a rigorous interpretation of the industrial application standard by requiring the disclosure of a more than speculative function of the invention at the time of the patent application.

Adapting law to the characteristics or needs of a particular industry is explained by the fact that, in theory, statutory patent law does not distinguish between different industries but provides technology-neutral protection to all kinds of technologies.[75] Since changes in an industry over time present significant structural problems for patent law, it has been argued that the unified rules suitable for the old, homogenous world are no longer appropriate in today's increasingly complex innovative landscape.[76] Any new technology presents the patent system with transitional difficulties in adjusting the standards of patentability, establishing doctrinal procedures and developing legal doctrine to address technology-specific aspects.[77] Since the EPC and other previous legislations were adopted, technology, and especially life sciences, has evolved dramatically over time. However, in general-purpose patent law regimes, rights cannot be modified according to the protection actually needed and the reward structure cannot be adapted according to the market structure in which the innovator operates.[78] Nonetheless, one size does not fit all, so adapting certain rules according to the needs of a particular industry might allow a better accommodation of scientific and technological developments as well as provide more optimal incentives to innovation.

101; Dan L Burk and Mark A Lemley, 'Policy Levers in Patent Law' (2003) 89 Virginia Law Review, 1575; Matthew Rimmer, *Intellectual Property and Biotechnology: Biological Inventions* (Edward Elgar 2008) 5–7. Nonetheless, a number of factors caution against explicitly tailoring patent systems to the needs of particular industries. In this regard, Article 27 of the TRIPS Agreement specifically prohibits Member States from discriminating in the grant of patents based on the type of technology.

[75] Dan L Burk and Mark A Lemley, 'Is Patent Law Technology-Specific?' (2002) 17 BTLJ 1155. See also Dan L Burk and Mark A Lemley, 'Tailoring Patents to Different Industries' in Emanuela Arezzo and Gustavo Ghidini (eds), *Biotechnology and Software Patent Law: A Comparative Review of New Developments* (Edward Elgar 2011).

[76] Burk and Lemley, 'Is Patent Law Technology-Specific?' (n 75).

[77] Cockburn (n 21) 115.

[78] Scotchmer, *Innovation and Incentives* (n 53) 117–118.

Despite the significant amount of patent applications over genetic material, in 1998 no patents had yet been granted for ESTs or SNPs in Europe.[79] However, it was the boom of patent applications concerning human gene sequences that triggered fears that patents would be granted over discoveries where the practical application was not yet determined.[80] The rapid developments that were taking place in the field of human genetics, coupled with the growing number of patent claims directed at obtaining patent protection over isolated gene fragments of unknown or highly speculative function raised the question of whether the existing patent system was sufficient or even appropriate to deal with these kinds of inventions.[81] To address this issue, it was argued that the requirement of industrial application should be more stringently applied for these types of claims,[82] which would help to ensure that inventions rather than fundamental discoveries were the subject of exclusive rights.

Therefore, through the Biotech Directive the Commission sought, among other objectives, to give greater importance to the criterion of industrial application in order to prevent the systematic grant of patents over gene fragments of unknown practical use. This approach also reflected the increased importance in the US of the corresponding requirement of utility.[83] More importantly, by inserting a strict interpretation of the requirement of industrial application for human genes, the framers of the Biotech Directive were also emphasizing that in cases such as the 1991 US NIH application for partial DNA sequences (ESTs) of which no function or utility was known at the time of the patent application, patent protection will not be awarded.[84] This idea received further support from the USPTO and the Japan Patent Office (JPO) at the

---

[79] Sven J R Bostyn, *Enabling Biotechnological Inventions in Europe and the United States: A Study of the Patentability of Proteins and DNA Sequences with Special Emphasis on the Disclosure Requirement* (European Patent Office 2001) 218.

[80] See Sven J R Bostyn, 'A Decade After the Birth of the Biotech Directive: Was It Worth the Trouble?' in Emanuela Arezzo and Gustavo Ghidini (eds), *Biotechnology and Software Patent Law* (Edward Elgar 2011) 224.

[81] Westerlund (n 31) 6.

[82] Trevor Cook, *Pharmaceuticals, Biotechnology and the Law* (2nd edn, Lexis Nexis 2009) 164.

[83] Rob J Aerts, 'Biotechnological Patents in Europe – Functions of Recombinant DNA and Expressed Protein and Satisfaction of the Industrial Applicability Requirement' (2008) 39 IIC 282.

[84] See Bostyn, 'A Decade After the Birth of the Biotech Directive: Was It Worth the Trouble?' (n 80) 224.

tripartite meetings,[85] and from other organizations and the scientific community,[86] who had expressed fears that patenting genes of unknown function could pose undue barriers to scientific research and ultimately impair innovation.

Nonetheless, to ensure that the genetics industry's high path progress continues, existing patent law doctrines, and in particular the industrial application criterion for patentability, must be carefully construed and enforced. Asserting the patent law requirement of industrial application depends on continuing agreement with respect to the policy that best promotes innovation, and the balance of competition and compensation in this field.[87] It should impose a real but not unbridgeable hurdle to the patentability of gene-related inventions,[88] thus addressing the interests of patentees and third parties and maintaining a fair basis for exclusivity that accurately reflects an invention's contribution to the art.[89] It is argued that patent law and policy must strike a balance between the traditional scientific aim of maximizing the production of knowledge and the traditional commercial aim of maximizing its exploitation.[90] In this regard, a well-thought-out and well-defined standard of industrial applicability may be the best means to achieve an appropriate and reasonably predictable balance between the private and public interests at stake in this industry.

4.3.2 Background and Adoption of Article 5(3) of the Biotech Directive

Despite the lack of unanimous acceptance that biological inventions could be patent eligible, the patent laws of European countries did not formerly include any special provisions regarding these types of inventions.[91] In this regard, the First Directive Proposal noted that:

---

[85] Cook, *Pharmaceuticals, Biotechnology and the Law* (n 82) 166.

[86] TD Kiley, 'Patents on Random Complementary DNA Fragments?' (1992) 257 Science 915; Michael A Heller and Rebecca S Eisenberg, 'Can Patents Deter Innovation? The Anticommons in Biomedical Research' (1998) 280 Science 698; Aerts, 'Biotechnological Patents in Europe – Functions of Recombinant DNA and Expressed Protein and Satisfaction of the Industrial Applicability Requirement' (n 83).

[87] Kenneth J Burchfiel, *Biotechnology and the Federal Circuit* (BNA Books 1995) 51–52.

[88] See Golden (n 74).

[89] See Westerlund (n 31) 89.

[90] Golden (n 74).

[91] Auth (n 60); Westerlund (n 31) 2.

> the patent system, when applied to biotechnology, encounters a number of particular problems ... in relation to ethics as well as in relation to the traditional patent law concepts of patentable subject matter, discovery, novelty, sufficient written disclosure, industrial applicability and the extent and exhaustion of patent protection.[92]

Therefore, as noted in Chapter 2, the First Directive Proposal aimed to reduce several of the deficiencies and discrepancies that statutory law such as the EPC had not addressed. In particular, although the original text proposed did not include any specific provisions regarding industrial application, subsequent proposals sought to address the fact that a growing number of patent applications concerning human genes were incapable of showing any industrial application.

It was not until the 1996 Amended Proposal that specific provisions regarding the industrial applicability of gene sequences were incorporated.[93] In response to the rejection of the 1992 Amended Proposal by the Parliament in 1995,[94] the Commission presented a more comprehensive text, which later constituted the basis for the final Biotech Directive. Although the Parliament's rejection was concerned with the proposal's insufficiency in addressing ethical issues, in the 1996 Amended Proposal the Commission also introduced changes regarding the industrial application of inventions relating to the human genes. In particular, Recital 15 of the 1996 Amended Proposal, which corresponded with Article 3 of the same text, specified that inventions concerning the human body and its parts were not patentable unless they were isolated and capable of industrial application.

Subsequently, Amendment 16 (regarding Recital 16 (new Recital 16e)) of the Report on the Proposal for a European Parliament and Council Directive on the legal protection of biotechnological inventions of 25 June 1997[95] introduced the requirement that at least one industrial

---

[92] Commission of the European Communities proposal for a Council Directive on the legal protection of biotechnological inventions (1988) [COM(88) 496 final – SYN 159] OJ C10/3, 7.

[93] Commission proposal for a European Parliament and Council Directive on the legal protection of biotechnological inventions (1996) [COM(95) 661 final] OJ C296/4. See also *Chiron Corporation v Murex Diagnostics Ltd* [1996] RPC 535.

[94] Decision on the joint text approved by the Conciliation Committee for a European Parliament and Council Directive on the legal protection of biotechnological inventions [1995] OJ C68/26.

[95] Report on the proposal for a European Parliament and Council Directive on the legal protection of biotechnological inventions (1997) [COM(95) 0661 –

application should be clearly disclosed in the patent application of inventions concerning sequences or partial sequences of a gene. The 1997 Report also proposed a similar amendment for Article 3 of the 1996 text by introducing a specific provision, Article 3(3), which stated that the industrial application of a sequence or a partial sequence of a human gene must be disclosed in the patent application.

Both the 1997 Amended Proposal[96] and the text finally adopted in 1998 included the provisions regarding the need to disclose the industrial application of inventions concerning human genes and gene sequences in the patent application. As a result, one area where the Biotech Directive apparently goes beyond previously existing requirements is Article 5(3), which mandates that in order to be patentable, the industrial application of an invention regarding the sequence or partial sequence of a gene must be disclosed in the patent application.

Article 5 of the Biotech Directive provides that the human body, at the various stages of its formation and development, and the simple discovery of one of its elements, including the sequence or partial sequence of a gene, cannot constitute patentable inventions. However, the Biotech Directive accepts in paragraph 3 of the same provision that such elements can be patented if, in addition to being isolated from the human body, applicants disclose the industrial application of these inventions. Recitals 22, 23 and 24 of the Biotech Directive further clarify this.

In particular, the Biotech Directive clarifies in Recital 23 that a mere DNA sequence without indication of a function does not contain any technical information and is not therefore a patentable invention.[97] Then Recital 24 provides that in cases where a sequence or partial sequence of a gene is used to produce a protein or part of a protein, it is necessary, in order to show industrial applicability, to specify which protein or part of a protein is produced or what function it performs in order to be patentable. Nonetheless, these recitals are somewhat confusing in the sense that proteins can have one or more biochemical functions (biological, chemical, pharmaceutical) and the Biotech Directive does not specify the type of function that has to be disclosed, for instance, the

---

C4-0063/96 – 95/0350(COD)] (Parliament Report 1997). See also *Chiron Corporation* (n 93) where the English Court of Appeal took the same approach towards the industrial applicability of the claimed polypeptides.

[96] Amended Proposal for a European Parliament and Council Directive on the legal protection of biotechnological inventions (1997) [COM(97) 446 final].

[97] Recital 23 recognizes that being industrially applicable is a precondition for an isolated discovery to acquire technical character (see Chapter 5).

commercial application resulting from the biochemical functions. Subsequent interpretation of this requirement by the EPO has provided clarification on this (see section 4.4).

4.3.2.1 Public response to the Biotech Directive's approach towards the industrial application of human gene sequences

An explicit provision requiring that applicants claiming a DNA sequence should specify in the patent application its industrial utility, which protein (or part of a protein) is produced or what function it performs was something that patent applicants were keen to avoid.[98] For example, a frequent argument against such an approach is that the commercial utility of DNA sequences often goes far beyond simple protein expression.[99] Besides, Article 5(3) and accompanying recitals received severe criticism from entities such as the British Group of the International Association for the Protection of Intellectual Property (AIPPI), who contended that to seek a specific function goes beyond the mandate of the Biotech Directive and that the same fairly low level of utility (or industrial applicability) that is required in other areas of technology should be applied to ESTs, SNPs and genomes.[100] It was claimed that raising this standard would interfere with the innovation progress in biotech sciences by hurting leading companies that had applied for patents on large numbers of ESTs, such as Incyte of Palo Alto, California or Human Genome Sciences, based in Rockville, Maryland.[101] However, this initiative has been widely praised by institutions like HUGO, which welcomed the Biotech Directive's clarifications on the interpretation of specific patentability requirements with regard to genetic inventions, in particular the provision that a mere DNA molecule and its sequence, without indication of function, does not contain technical information and cannot be an invention.[102] HUGO noted that while the USPTO accepted the use of tags as probes to identify specific DNA sequences as demonstrating utility, this decision would affect later innovation in which ESTs are involved since it would be in one way or another dependent on such

---

[98] Claire Baldock and Oliver Kingsbury, 'Where Did It Come From and Where Is It Going? The Biotechnology Directive and Its Relation to the EPC' (2000) 19 Biotechnology Law Report 7.

[99] Ibid.

[100] Claire Baldock and others, 'Report Q 150: Patentability Requirements and Scope of Protection of Expressed Sequence Tags (ESTs), Single Nucleotide Polymorphisms (SNPs) and Entire Genomes' (2000) 22 EIPR 39.

[101] Enserink (n 34).

[102] HUGO 2000 Statement (n 49).

patents. The organization emphasized that DNA molecules and their sequences, be they full-length, genomic or cDNA, ESTs, SNPs or even whole genomes of pathogenic organisms, if of unknown function or utility, as a matter of policy, should be viewed in principle as part of pre-competitive information, and initiatives to put them in the public domain, as the Consortium of industry and academia attempted, are desirable.[103] According to HUGO the process required to isolate the full-length gene and determine its biological function from an EST is considerably more difficult, produces more social benefit and thus warrants the most protection and incentive.[104]

The UK Chartered Institute of Patent Attorneys (CIPA) also expressed its support for the Biotech Directive's approach as a means for striking a balance between innovators and society by ensuring that speculative patent applications relating to human genes will not succeed.[105] Further, the leader of the US component of the HGP, Francis Collins, praised the similar US initiative since it pursued:

- reducing the number of 'generation one' patents, where there is just a sequence of unknown utility
- increasing the number of 'generation two' applications, where there is a sequence and homology suggests a function and ultimately
- 'generation three' inventions, that are more suitable for patent protection because their gene sequences have a biochemical, cell biological or genetic data describing function.[106]

4.3.3 Incorporation of Article 5(3) into the EPC and National Laws

The EU is increasingly participating in patent policy and after the incorporation of key provisions of the Biotech Directive into the text of the EPC, this part of EU law is now an essential component of European patent law regarding the protection of biotechnological inventions. However, this does not give the EU jurisdiction over the EPC since the EPO is

[103] Ibid.

[104] Melanie J Howlett and Andrew F Christie, 'An Analysis of the Approach of the European, Japanese and United States Patent Offices to Patenting Partial DNA Sequences (ESTs)' (2003) 34 International Review of Industrial Property and Copyright 581.

[105] The Chartered Institute of Patent Attorneys (CIPA), 'Patenting of Human Gene Sequences' (April 1994 (last revised June 2006)) <available at: http://www.cipa.org.uk/pages/info-papers-human> accessed 27 July 2013.

[106] Rimmer (n 74) 144–147.

not a EU institution and the EPC is thus not part of EU law.[107] However, the EPO has expressly indicated its intention of using the Biotech Directive as a supplementary means of interpretation with regard to the rules reproducing the Biotech Directive's text (see section 3.2.3).[108]

In 1999, by a decision of the EPO's Administrative Council,[109] some of the Biotech Directive's provisions were incorporated into the Implementing Regulations to the EPC. Article 5 of the Biotech Directive was reproduced in Rule 23e (now Rule 29)[110] of the EPC Implementing Regulations and entered into force, as did the rest of the incorporated provisions on 1 September 1999. The new rules were deemed to be retrospectively applicable to all applications awaiting grant, cases and opposition proceedings that were pending at that date.[111]

Since the Biotech Directive is part of EU legislation, all EU Member States were required to incorporate the Biotech Directive's provisions into their national laws. With regard to Article 5(3) and related recitals, while most countries have adopted these provisions, some have diverged in implementing them. For example France incorporated the Biotech Directive in the Intellectual Property Code and excluded from patentability discoveries of components of the human body like whole or partial human gene sequences, however it has not expressly provided for the patentability of isolated gene sequences that disclose their industrial application.[112] Another example is Denmark, where the Biotech Directive was incorporated into Danish Law in May 2000 but Article 5(3) was

---

[107] Convention on the Grant of European Patents (European Patent Convention (EPC)) of 5 October 1973, art 4.

[108] Ibid, rule 23b (now rule 26(1)).

[109] EPO Decision of the Administrative Council of 16 June 1999 amending the Implementing Regulations to the European Patent Convention (OJ EPO 1999, 437).

[110] The Implementing Regulations to the EPC of October 1973 were amended by the Decision of the Administrative Council of the European Patent Organisation of 9 December 2004 amending the Implementing Regulations to the European Patent Convention and the Rules relating to fees (OJ EPO 2004, 11). The revised version of the EPC Implementing Regulations entered into force on 13 December 2007.

[111] *Transgenic animals/HARVARD* (T 0315/03) [2004] (EPO (TBA)) point 5; *Use of embryos/WARF* (G 0002/06) [2008] (EPO (EBA)) points 13 and 14.

[112] French Intellectual Property Code of 1992, art L611-18 (inserted by Act No 2004-800 of 6 August 2004, Article 17a II, Official Journal of 7 August 2004), see Denis Schertenleib, 'An Introduction to European Intellectual Property Rights in Intellectual Property' in Paul England (ed), *Intellectual Property in the Life Sciences: A Global Guide to Rights and their Applications* (Globe Law and Business 2011). See also CIPA, 'Patenting of Human Gene Sequences' (n 105).

reproduced in Section 18(1)(5) of the Danish Patent and Supplementary Protection Certificate Order,[113] which requires an 'express explanation' of the industrial application of a gene. The Biotech Directive's essential requirement of a specific definition of the industrial applicability of a gene and corresponding protein is thus not mentioned in the Danish Patents Act itself, but only in this complementary Order. However, these cases are exceptions to the majority of countries, which have implemented Article 5 of the Biotech Directive (Rule 29 of the EPC) entirely as Members of the EU and signatory countries of the EPC.

4.3.3.1 Status of new Rule 29(3) EPC

After the incorporation of Article 5(3) of the Biotech Directive into the EPC Implementing Regulations, questions arose regarding the status of the new requirement. First, clarification was needed regarding the status of EPC Rule 23(e)(3) (now Rule 29(3)) with respect to EPC's Article 57 general requirement of industrial application. Since the EPO is not a body of the EU, there were doubts as to the relationship between the Biotech Directive's provisions, which are rules of EU supranational law, and the EPC rules, which emanate from an intergovernmental treaty. According to Article 249 of the Treaty establishing the European Community (EC Treaty),[114] the provisions of the Biotech Directive are binding to all the Member States. However, the same provisions, after being incorporated into the EPC Implementing Regulations, are under Article 164(2) of the EPC subsidiary to the provisions of the EPC, and thus to Article 57 of the EPC in the event of conflict.

The EPO had the opportunity to clarify this issue in Icos Corporation where the Opposition Division (OD) rejected the idea of superiority of Article 57 of the EPC.[115] However, an appeal to this decision was declared inadmissible and the EPO Board of Appeal could not discuss the issue further.[116] Later in Zymogenetics the appellant argued that Rule

[113] Danish Patent and Supplementary Protection Certificate Order No 93/2009; Kasper Frahm and Sture Rygaard, 'An Introduction to European Intellectual Property Rights in Intellectual Property' in Paul England (ed), *Intellectual Property in the Life Sciences: A Global Guide to Rights and their Applications* (Globe Law and Business 2011).

[114] Consolidated version of the Treaty establishing the European Community of 29 December 2006 [2006] OJ C321 E/37.

[115] *Novel V28 Seven Transmembrane Receptor/ICOS CORPORATION* [2002] OJEPO 293 (EPO (OD)).

[116] *V28 receptor/ICOS* (T 1191/01) [2002] (EPO (TBA)). For a further discussion on the implications of the decision in Icos Corporation with regard to

23(e)(3) could not change the interpretation of Article 57 EPC since the Implementing Regulations were not to be used for changing the initial basic meaning of the EPC without infringing Article 164(2) EPC. According to the appellant, Recitals 22, 28 and 34 of the Biotech Directive all emphasized and made clear that the legislative intention was not to change relevant basic law, including the provisions on susceptibility to industrial application. In this regard, requiring an indication of a use or a function of a DNA sequence would be in conflict with the EPC since such a requirement was not derivable from Article 57 itself, and the later should prevail. However, the Examining Division found no conflict between Rule 23(e)(3) and Article 57 of the EPC.[117] It stated that in the light of the new rule, Article 57 EPC could no longer be interpreted in the classical sense, according to which if the protein in question could be made then industrial applicability was shown; but the existence of Rule 23(e)(3) EPC required an examination as to whether or not the requirement of Article 57 EPC was fulfilled.

A second question that needed clarification was whether decisions concerning the interpretation of Rule 29 could be appealed before the CJEU. In Warf[118] it was argued that since Rule 29 mirrors the wording of Article 5 of the Biotech Directive, the EPO EBA in interpreting Rule 29 EPC is interpreting the law of the EU and is thus required (as a court or tribunal of a Member State against whose decision there is no judicial remedy) by Article 234 of the EC Treaty to ask for a ruling by the CJEU. However, the EBA held that because the Boards of Appeal of the EPO are not courts or tribunals of a Member State of the EU and they apply the provision because it is law under a specific rule of the Implementing Regulations to the EPC, and not because the Biotech Directive is a source of law to be applied directly, there is no power under the EPC for a Board of Appeal to refer questions to the CJEU. Therefore, only national courts could refer questions of law to the EU Court of Justice.

---

the status of Rule 23(e)(3) see Aerts, 'The Industrial Applicability and Utility Requirements for the Patenting of Genomic Inventions: A Comparison between European and US Law' (n 33).

[117] *Haematopoietic cytokine receptor/ZYMOGENETICS* (T 0898/05) [2006] (EPO (TBA)) point VII (summary of the reasons given by the examining division).

[118] *Warf* (n 111) points 4, 5 and 6.

4.3.4 The Initial Interpretation of the New Standard by the EPO

Article 5(3) of the Biotech Directive did not attract much attention during and immediately after the drafting and negotiation process.[119] Besides, while the Biotech Directive made clear how important the requirement of industrial application was in gene patenting, it did not provide concrete indications on how the EPO and national patent offices were to apply the standards set in Article 5(3) and related recitals. Thus given the low number of EPO and national courts' decisions dealing with the issue of industrial applicability, it remained uncertain how the relevant authorities should interpret the new provisions. The most certain feature of Article 5(3) was that the criterion of industrial application when applied to gene fragments, and the proteins they code for, was no longer compatible with the traditional broad interpretation of Article 57 EPC.

In 2000, comparative studies between the Japanese, US and European patent offices shed light on the interpretation of industrial application in the patentability of DNA fragments.[120] In the examples discussed, the EPO considers two interpretations of Article 57 and distinguishes between cases where a polynucleotide can clearly be made, and thus the utility requirement is fulfilled, and cases where the usefulness of an EST is not known and thus no one would make it in an industrial context. In the last case, the EPO concludes that if an EST has no specific function, it cannot be used in industry and therefore does not satisfy the utility requirement.

The trilateral study also contained a conclusion with the main points revealed through the comparative exercise, which could be summarized as follows:

- a mere DNA sequence without indication of a function or specific asserted utility is not a patentable invention
- a DNA sequence, of which specific utility is disclosed such as use as a probe to diagnose a specific disease, is a patentable invention as long as there are no other reasons for rejection and
- a DNA sequence showing no unexpected effect, obtained by conventional methods, which is assumed to be part of a certain

---

[119] Bostyn, 'A Decade After the Birth of the Biotech Directive: Was It Worth the Trouble?' (n 80).

[120] EPO, JPO and USPTO, Trilateral Project B3b, Comparative study on biotechnology patent practices – Theme: Patentability of DNA fragments (Trilateral offices 2000).

structural gene based on its high homology with a known DNA encoding protein with a known function, is not a patentable invention.

The three offices, after discussion at the Trilateral Technical Meeting of June 2000, adopted two additional conclusions:

- all nucleic acid molecule related inventions, including full-length cDNAs and SNPs, without indication of function or specific, substantial and credible utility, do not satisfy industrial applicability, enablement or written description requirements and
- isolated and purified nucleic acid molecule related inventions, including full-length cDNAs and SNPs, of which a function or specific, substantial and credible utility is disclosed, and which satisfy industrial applicability, enablement, definiteness and written description requirements, would be patentable as long as there is no prior art (novelty and inventive step) or other reasons for rejection such as best mode in the US or ethical grounds in the EPO or JPO.

A particular question that arises from these conclusions is whether the EPO equates the concepts of 'function', 'utility' and 'industrial application' within the context of DNA patenting. From the trilateral study it seems that the EPO refers indistinctively to the three terms.[121] In contrast, the Biotech Directive does not use the term 'utility' and, with regard to the use of the terms 'function' and 'industrial application', different parts of the Biotech Directive seem to offer different insights.[122] Recital 24 of the Biotech Directive appears to refer to the specification of the coded protein's function as the means to show industrial applicability wherein Article 5(3) refers to the industrial application of the protein-encoding gene sequence. Recital 23 then uses the word 'function' on its own and thus can be interpreted both ways. Nonetheless, a link can be established between the function related to the gene sequence on the one hand and the industrial application connected to it on the other hand. Besides, the history of the origins of Article 5(3) shows a close relationship between both terms. While the wording of Article 5(3)

[121] Note here that the EPO, JPO and USPTO in the tripartite meetings (n 120) use the terms 'utility', 'function' and 'industrial applicability' to refer to the same idea, thus suggesting that the three terms are interchangeable (at least within the context of DNA inventions).

[122] See Schertenleib, 'The Patentability and Protection of DNA-Based Inventions in the EPO and the European Union' (n 12).

adopted by the Parliament in the first reading still referred to 'function',[123] the wording as finally adopted refers to the 'industrial application' of the sequence or partial sequence of a gene. Furthermore, in any event, what is certainly clear is that the Biotech Directive sought to avoid the grant of patents to inventions that, although they can be manufactured, do not have any known practical applicability at the time of the application.

4.4 THE EPO CASE LAW ON RULE 29(3) OF THE EPC

In addition to the 2000 comparative study, after incorporating Article 5(3) of the Biotech Directive into the EPC Implementing Regulations, the EPO has developed a substantial body of case law addressing the question of industrial application in human gene patents pursuant to Rule 29(3) of the EPC.

The standard set by Rule 29(3) of the EPC raised a number of questions among applicants and other stakeholders regarding how the new rule would be applied in practice by the EPO boards. First, given the influence of the US approach towards the utility of biotech inventions in the discussions prior to the adoption of Article 5(3) of the Biotech Directive, it was expected that the US utility doctrine could have a role in the EPO's interpretation of Rule 29(3). However, no EPO decisions were available to confirm such a presumption. A second question related to where the EPO would draw the line between abstract interesting research results and industrially applicable inventions. Specifically, clarification was needed on whether the EPO would follow previous case law, and it would therefore be enough that an invention could be made or used to comply with the industrial applicability criterion. If however more than that had to be shown, then a third question would arise regarding the type of information that has to be disclosed in the patent application. In this regard, given the importance of genetic inventions to society, the EPO would also need to consider whether there would be exceptions to the standards established by the EPO case law on Rule 29(3), for example for inventions that are important for healthcare. Finally, clarification regarding the type and amount of evidence that is acceptable for purposes of Rule 29(3) was also highly desirable. The EPO has gradually answered these questions by discussing each of those issues through a number of decisions on the interpretation of Rule 29(3) EPC.

---

[123] 1997 Amended Proposal (n 96). See also Oser (n 18).

4.4.1 Following the US Approach: Specific, Substantial and Credible Utility

The first time that the EPO gave a decision dealing with the industrial applicability of DNA inventions was in the Icos Corporation case,[124] where the patent related to a human DNA sequence encoding the amino acid sequence of the polypeptide V28 seven transmembrane receptor (V28 7TMR).[125] Opponents argued that no practical use of the invention was shown while the patentee held that the specification disclosed how to make the V28 protein and also the uses of this protein, mainly as a receptor involved in immunological processes.[126] In particular, the opposition held that the specification in this case did not disclose a ligand binding (connection) to said protein (V28 7TMR) and thus its function as a receptor was only based on the structural characterization of the claimed sequence. Therefore, immunological properties specific to V28 7TMR, for example other than general housekeeping cellular functions, could not be acknowledged either.[127]

The EPO OD found the patent invalid for lack of industrial application, inventive step, sufficiency and technical character. With regard to industrial application, this decision is noteworthy because when assessing the practical function of the claimed invention under Article 57 in conjunction with Rule 23b-e (now Rule 29(2) and (3)) of the EPC, the EPO followed the US 'specific, substantial and credible test' and found the list of potential uses disclosed in the application to be speculative in the sense that they were not specific (concrete), substantial and credible since the involvement of the claimed protein in immunological or inflammatory processes had not been demonstrated in vivo, and thus was not industrially applicable.[128] Since no function of the V28 protein was sufficiently disclosed in the patent application, for instance a biological

---

[124] *Icos Corporation* (n 115).

[125] Seven transmembrane receptors are able to pass information seven times through the cell membrane. Such transmission is initiated by internal signal transduction pathways that are activated after detecting the presence of molecules outside the cell.

[126] *Icos Corporation* (n 115) point 8(ii).

[127] Ibid, point 2(ii).

[128] Ibid, point 9(i). See also; T J Kowalski, A Maschio and S H Megerditchian, 'Dominating Global Intellectual Property: Overview of Patentability in the USA, Europe and Japan' (2003) 9 Journal of Commercial Biotechnology 305; Fitt and Nodder, 'Specific, Substantial and Credible? A New Test for Gene Patents' (n 21).

function valuable for therapeutic use or as a marker for use in diagnostics, it was found that no industry would be interested in manufacturing such protein.

Icos Corporation is the leading case on the interpretation of industrial application for human gene-related inventions. This decision shows that for human DNA inventions to meet the requirement of industrial applicability, it is no longer enough that the claimed invention can be made or used, but a substantial, credible, specific and therefore not speculative use of the invention has to be disclosed at the time of the patent application. The OD at the EPO interpreted the Biotech Directive's requirement of disclosing the industrial application of human gene patents in line with the US utility doctrine formulated in the USPTO Utility Examination Guidelines.[129] Thus, this decision again highlights the increasingly close and interchangeable identity between the definitions of industrial application and utility.[130] Furthermore, it also confirms the shift in the EPO's interpretation of industrial application, from a requirement of being made or used to a requirement that demands an explicit specification of usefulness.

4.4.2 The Distinction between Interesting Research Results and Industrially Applicable Inventions

Although the decision in Icos Corporation established a landmark ruling for future decisions on the industrial application of human gene inventions, it left open the question of whether a research use alone can amount to industrial applicability. Even though the first opponent in this case (Smithkline Beecham) specifically argued that the use of a newly identified protein in research is not equivalent to industrial application,[131] the OD decided not to discuss this issue and did not provide any guidance on the matter. However, later in Salk Institute[132] the EPO TBA

---

[129] United States Patent and Trademark Office (USPTO), Utility Examination Guidelines (2001). The UK Guidelines for biotechnology also incorporate the US substantial, specific and credible test as a means of identifying the useful and practical purpose of a claimed invention.

[130] See Sivaramjani Thambisetty, 'Legal Transplants in Patent Law: Why Utility is the New Industrial Applicability' Law, Society and Economy Working Papers 6/2008 (2009) 49 Jurimetrics J 155. See also patent BL O/286/05, *Aeomica Inc* (2005) where the UK Intellectual Property Office (UKIPO) applied the US specific, substantial and credible to reject the patent application for lack of industrial application.

[131] *Icos Corporation* (n 115) point 8(i). See also Thambisetty (n 130).

[132] *Multimeric Receptors/SALK INSTITUTE* (T338/00) [2002] (EPO (TBA)).

took the opportunity to discuss this question and concluded that interesting research results without a readily identified industrial application would prove insufficient. In this case, the board felt it necessary to examine whether, for the claimed heterodimeric human receptor or dimer[133] and for the claimed method to modulate transcription activation of a gene, which is the first step in gene expression, the way in which they are capable of being exploited in industry could be ascertained from the description; or whether what the applicant had described was merely an interesting research result that might yield a yet to be identified industrial application.

In answering this question, the board agreed with the appellant in that the application disclosed the presence of cooperative interactions, that is, a stable association or affinity between two specific types of receptors, to form heterodimeric receptors and also provided evidence on the use of these heterodimers for modulating suitable transcription expression systems.[134] According to the TBA, the activities and products disclosed in the application were not aimed at an abstract or intellectual character but at a direct technical result that could be clearly applied in an industrial activity like modulation of the expression of a gene or product of interest in a particular expression system, screening of products with specific pharmacological activity, etc.

A few years later in the Max Planck case,[135] the TBA further clarified where the borderline between real inventions and mere research results resides. In this case the application disclosed a newly identified, isolated, enriched and purified polypeptide of human origin called brain-derived phosphatase 1 (BDP-1), which was described as having certain structural features involved in cellular signal transduction pathways and a possible role in the maintenance of basic cellular functions (cellular housekeeping) and in certain types of cancer. The description also suggested that the claimed protein was a member of a specific protein family (the protein tyrosine phosphatase proline-, glutamic acid-, serine- and threonine-rich (PTPases-PEST) family). BDP-1 was the first non-transmembrane protein that is not capable of passing across a membrane, for which this combination of distinct structural and functional features was disclosed, making it a promising target for therapy intervention, the manufacture of anti-cancer drugs and for the elucidation of the molecular

[133] A dimer is a molecule consisting of two identical simpler molecules.

[134] *Multimeric Receptors* (n 132) point 3. See also Fitt and Nodder, 'Specific, Substantial and Credible? A New Test for Gene Patents' (n 21); Leverve (n 54).

[135] *BDP1 Phosphatase/MAX PLANCK* (T 0870/04) [2005] (EPO (TBA)).

mechanisms underlying cancer development. However, the EPO Examining Division (ED) found the patent invalid for lack of industrial application. The decision was upheld by the TBA against the applicant argument that the elucidation of such a biological effect and/or cellular function should provide sufficient to justify an industrial application.

In its decision, the TBA emphasized that the fact that a substance could be produced or made in some way is not enough, but that a profitable use for such substance needs to be disclosed.[136] Besides, the TBA relied on public policy grounds to reject the application by noting that the purpose of granting a patent is not to reserve an unexplored field of research for an applicant,[137] and thus research results, although they can be very interesting scientific achievements of remarkable merit, are not eligible for patent protection unless it is shown that such results can be applied industrially.[138] This is because the whole burden to guess or find a manner to exploit an invention cannot be left to the reader, who would not have known if any practical application actually exists.[139] Therefore, the applicant has the duty to identify and describe the practical value of the claimed invention and how such invention could be used in industry.

In view of the decisions in the cases of Salk Institute and Max Planck, it appears that the EPO has established a distinction between inventions that show some 'profitable use', and are thus susceptible of industrial application, and patent applications consisting of vague and speculative indications of potential functions that amount to merely interesting research results that require further work to identify a specific industrial application. Reflecting this approach, the decision in Zymogenetics developed several principles that further defined the concept of 'profitable use' set in Max Planck. In this case the application disclosed an isolated polynucleotide, encoding a ligand-binding receptor polypeptide[140] comprising a sequence of amino acids of the human transmembrane receptor Zcytcor1. It contained claims to the polynucleotide, the polypeptide and to antibodies that specifically bound to the polypeptide. The receptor was identified as a putative member of a family of cytokine

---

[136] Ibid, point 4. See also Robert Fitt and Edward Nodder, 'The Industrial Applicability of Biotechnology Patents – A New Test for Europe' (2009) 28 Biotechnology Law Report 151; Andrew Sharples and Robert L Smith, 'Patents Claiming Genetic Sequences' (2009) 216 Patent World 32.

[137] *Max Plank* (n 135) point 21.

[138] Ibid, point 6. See also Fitt and Nodder, 'Specific, Substantial and Credible? A New Test for Gene Patents' (n 21); Spencer (n 26).

[139] *Max Plank* (n 135) point 19.

[140] A type of protein that binds to another entity to form a larger complex.

receptors[141] known as the haematopoietin receptor family, which had a role in the proliferation, differentiation and activation of immune cells. Furthermore, therapeutic applications directly derivable from the biological function of the invention had been adequately disclosed. They included the treatment of diseases such as rheumatoid arthritis, multiple sclerosis and diabetes, and a reduction in the rejection of tissue or organ transplants and grafts. The ED found the patent invalid for lack of industrial application under Article 57 and Rule 23(e)(3) EPC. However, on appeal, the TBA held that the application had a plausible practical utility.

In this case, the board considered that the need to show a profitable use was to be understood in a broader sense than that of an actual or potential economic profit or of a commercial interest. The invention claimed should have a sound and concrete technical basis that could lead a skilled person, without the need for carrying out further research,[142] to an immediate practical exploitation in industry.[143] In this regard, 'concrete benefit' refers to the need of disclosing in definite technical terms the purpose of the invention and how it can be used in industrial practice to solve a given technical problem, this being the actual benefit or advantage of exploiting the invention. On the other hand, 'immediate' conveys the need for this to be derivable directly from the description, if it is not already obvious from the nature of the invention or from the background art.[144] Purely theoretical or speculative claims would not be acceptable under this interpretative standard.

Subsequent EPO decisions followed the interpretation of profitable use that was set in Zymogenetics. For example, in Bayer the TBA found that the only use of the claimed human polypeptide comprising the amino acid sequence of sequence SEQ ID NO: 24 was to find out more about the polypeptide itself and its natural function(s), and thus it did not amount to an immediate concrete benefit.[145]

---

[141] The insufficiency of cytokine receptors in the body has been associated with certain debilitating immunodeficiency conditions.

[142] *Zymogenetics* (n 117) points 6 and 7. See also Fitt and Nodder, 'Specific, Substantial and Credible? A New Test for Gene Patents' (n 21); Spencer (n 26).

[143] *Zymogenetics* (n 117) point 5.

[144] Ibid, point 6.

[145] *Serine Protease/BAYER* (T 1452/06) [2007] (EPO (TBA)) point 23. See also Fitt and Nodder, 'Specific, Substantial and Credible? A New Test for Gene Patents' (n 21); Sharples and Smith (n 136).

4.4.3 Types of Functions that Can Make a Protein Industrially Applicable

Pursuant to Rule 29(3) of the EPC, the function of the protein encoded by a claimed gene sequence has to be disclosed. However, this rule does not specify the exact type of function that needs to be disclosed for complying with this requirement. The TBA at the EPO addressed this question in Zymogenetics. In this case the TBA first explained that the function of a protein, and the nucleic acid encoding it, could be seen at three different levels, namely:

(1) molecular function (biochemical activity of the protein)
(2) cellular function (participation of the protein in cellular processes) and
(3) biological function (influence of those cellular processes in which the protein takes part within a multicellular organism, for example in the forms of cancer, inflammation, etc.).[146]

The TBA then clarified that each functional level may encompass multiple functions, but that in some cases the elucidation of one particular level of function might result, under certain conditions, in a straightforward industrial application, even if the other levels of activity remain completely unknown or only partially characterized. In addition, the TBA decision recognized that there can be cases where none of these levels is shown but the industrial applicability of a protein is derived from its specific structural features.[147] Moreover, in accordance with the broader interpretation of profitable use set in Max Planck, the TBA confirmed that none of those levels is more important than the others[148] and that the immediate concrete disclosure of one of them might prove sufficient for acknowledging industrial application.

4.4.4 Inventions with Functions Known to be Essential for Human Health Purposes

The TBA explained in Max Planck that in cases where the function of an invention is known to be essential for human health, such invention

---

[146] *Zymogenetics* (n 117) points 29 and 30. See also *GPCR-like receptor/ PHARMACIA* (T 0641/05) [2006] (EPO (TBA)) point 13.

[147] *Zymogenetics* (n 117) points 29 and 30. See also *Pharmacia* (n 146) point 13.

[148] *Zymogenetics* (n 117) point 30.

immediately suggests a practical application. These could be for instance the cases of insulin, human growth hormone or erythropoietin. However, in cases where a substance is isolated and characterized but the function is unknown, complex or incompletely understood, and no disease is identified as being attributable to an excess or deficiency of the substance, then the invention is deemed to have no practical application.[149] In other words, a protein related invention shows industrial applicability if:

- it is identified
- its function is completely understood
- a disease or condition is determined and
- its correlation with such disease or condition is characterized.

Moreover, it must be noted that in light of the Max Planck test for inventions relevant to human health, it would appear that the understanding (disclosure) of the biological function is of primary importance since it is there that a correlation with a disease or condition may be made. In contrast, knowledge of the two other functions, molecular and cellular, might be more suitable to characterize the nature of the relationship between the claimed substance and the disease or condition,[150] which is also required in the test.

Later in the Genentech case, the patent application provided a structural characterization of human polypeptide receptors, which enabled their assignment to the category of receptors that bind members of the PF4A family of chemokines;[151] and thus indicated what their function could be, for instance the disclosed receptors could have therapeutic potential in connection with inflammation and wound healing.[152] However, their ligands had not been characterized, and thus their function was not completely understood. The OD at the EPO rejected the claims for lack of a credible function. On appeal, taking into account the common general knowledge at the filing date, the board found that chemokines as a family were considered not only to be interesting in fundamental research, but it acknowledged that PF4-related proteins are attractive targets for the development of new therapeutic agents, for example for

---

[149] *Max Plank* (n 135) points 5 and 6. See also Sharples and Smith (n 136).

[150] Leverve (n 54).

[151] *PF4A receptors/GENENTECH* (T 0604/04) [2006] (EPO (TBA)). Chemokines are proteins that help cells to produce signals (communicate) to carry out basic cellular activities and coordinate cells' functions.

[152] Ibid, point 13.

anti-inflammatory, healing and tissue repair purposes, and are thus important for the pharmaceutical industry irrespective of whether or not their role had been clearly defined.[153] Because of its relevance to human health purposes, industrial applicability was therefore acknowledged.

The same approach was reiterated in Zymogenetics,[154] where the TBA recognized that if an invention is clearly described and plausibly shown to be of use in healthcare, for example to cure a rare or orphan disease, then the benefit is immediate and concrete even without the pursuit of trade.

4.4.5 Evidence: What Type and How Much is Required

The introduction of Rule 29(3) of the EPC also raised questions regarding the amount and type of evidence that is necessary to consider that a gene sequence and related protein show an indication of function, and is therefore a patentable invention. In this regard, the OD in Icos Corporation left this question open by limiting its decision to a general comment saying that despite the data provided, the polynucleotide sequences disclosed in the patent application were based on a speculative function. However, several later EPO decisions and statements have provided applicants with significant guidance on the type and amount of evidence that is required to show industrial application. Yet each case has to be considered on its own merit and if a profitable use can readily be identified based on the description, taking into account common general knowledge at the priority date as well as the prevalent view of the person skilled in the art, industrial application can be acknowledged despite the absence of actual experimental data.[155] Hence, an educated guess of a plausible function might be enough in some cases.[156] However, if the alleged function of a claimed substance is not credible beyond mere speculation, the EPO will request experimental evidence demonstrating the function.[157] Therefore, what is important is that the evidence shown is

---

[153] Ibid, points 17 and 18. See also Spencer (n 26).

[154] *Zymogenetics* (n 117) point 8.

[155] *Genentech* (n 151) point 14; *Zymogenetics* (n 117) point 20; *Human Delta3 – Notch/MILLENNIUM* (T 2101/09) [2013] (EPO (TBA)) point 13.

[156] *Zymogenetics* (n 117) point 27.

[157] EPO, JPO and USPTO, Trilateral Project B3b, Mutual understanding in search and examination – Comparative study on biotechnology patent practices – Theme: Nucleic acid molecule-related inventions whose functions are inferred based on homology search (Trilateral offices 2000).

enough to identify a function no matter how much, if any, experimental data are provided.

Another question for consideration relates to the type or nature of the evidence that is admitted to demonstrate industrial applicability. Today genetic research heavily relies on computer-assisted methods like bioinformatics, thus applicants would need to know whether *in silico* evidence is deemed sufficient or if wet-lab results are still required. In this regard, the EPO has confirmed in several decisions the validity of bioinformatics and other computer-assisted methods for assigning a function to a specific substance.[158] Although the probative value of such methods is to be seen on a case-by-case basis,[159] these decisions show that the EPO recognizes their value as an integral part of scientific investigations and their capacity to lead to plausible conclusions regarding the function of a product before it is actually tested.

Finally, the EPO has clarified the degree of homology that is needed in order to show sufficient probative value. In principle, the EPO will not accept homology data if the homology is below 55 per cent or if it involves only homology across a protein's restricted region or motif. However 80 per cent homology in DNA sequences and 95 per cent at the protein level could be acceptable if such a degree of homology is present across the whole coding sequence and not only in a motif.[160] The decisions in the Pharmacia and Human Delta cases illustrate such approach. In Pharmacia, a degree of 89.6 per cent amino acid sequence identity was rejected,[161] while in Human Delta, the TBA accepted a structural identity of at least 95 per cent as sufficient proof for the alleged function.[162] Moreover, in this case the TBA emphasized the fact that DNA claims are not required to be limited to a very specific sequence but

[158] *Zymogenetics* (n 117); *Bayer* (n 145) point 7. However, in *Pharmacia* (n 146) point 14, while the approach adopted by the TBA in Zymogenetics was not contested, it was suggested that *in silico* evidence might prove to have a less probative value.

[159] *Zymogenetics* (n 117) point 22.

[160] EPO, JPO and USPTO Trilateral Project B3b, Mutual understanding in search and examination – Comparative study on biotechnology patent practices – Theme: Nucleic acid molecule-related inventions whose functions are inferred based on homology search (n 157).

[161] *Pharmacia* (n 146) point 12. The board concluded that, although the CEGPCR1a clone (defined in the application as a splice variant of CEGPCR1 (AC7.1)) has some structural features of a G protein-coupled receptors (GPCRs) like receptor, it also presents some other non-conventional and unusual features. A degree of homology of 89.6 per cent is considered to be speculative.

[162] *Human Delta3-Notch* (n 155) point 16.

may embrace molecules having a certain degree of homology and/or identity to such specific sequence,[163] which allows applicants to protect their inventions against arbitrary modifications of the specific sequences.

4.4.6 Implications for Future Patent Applicants

A set of basic principles can be drawn from the EPO case law on the industrial application of gene sequences and related proteins.

(1) The EPO has confirmed the growing closeness between the US utility criterion and the European requirement of industrial applicability. The implementation of Rule 29(3) of the EPC has shed light on the question of whether the European patent law requirement of industrial application is actually equivalent to the US requirement of utility. In this regard, the EPO interpretation of the Biotech Directive's new rule on the industrial application of human DNA inventions indicates that both criteria, industrial application and utility, are essentially interchangeable, at least within the context of human gene patenting. In fact it seems that the requirement of industrial application as applied by the EPO has evolved toward an actual test of the usefulness of an invention,[164] closer to the US approach. The general requirement of industrial application would be fulfilled if the invention can be made or used while utility is fulfilled when the invention is useful and has a positive and practical benefit to society.[165] While the appropriateness of adopting the US specific, substantial and credible test for assessing the industrial applicability of European patents might be somewhat questionable,[166] Rule 29(3) of the EPC harmonizes both concepts by introducing a more 'useful' character for the European requirement.[167] In theory, this more utility-like approach is only for human DNA patents, but its implications have extended to other fields.

---

[163] Ibid, point 13.

[164] Leverve (n 54).

[165] Aerts, 'The Industrial Applicability and Utility Requirements for the Patenting of Genomic Inventions: A Comparison between European and US Law' (n 33).

[166] Thambisetty (n 130).

[167] Aerts, 'The Industrial Applicability and Utility Requirements for the Patenting of Genomic Inventions: A Comparison between European and US Law' (n 33).

(2) The case law on Rule 29(3) EPC has provided a clearer distinction between what would be mere research results and inventions with a real application in industry. On the one hand, there are those patent applications consisting of vague and speculative claims over inventions that do not disclose any practical utility, and thus further research would be needed to identify at least one potential use of such inventions in industry. And on the other hand, there are applications that describe a profitable, immediate and concrete benefit; that is, a substantial, specific and credible utility of the claimed invention.
(3) The EPO boards have also expressly recognized the importance of genetic research for biomedical research by exempting patent applications concerning inventions that are essential to human health from showing a specific application in industry. In these cases, the EPO accepts the idea that if an invention can lead to the development of important solutions for human healthcare, then industrial applicability can be immediately acknowledged.
(4) There is certain flexibility on the types of functions that can be disclosed for the purpose of showing industrial applicability. However, the appropriateness of disclosing a particular type of function is assessed on a case-by-case basis.
(5) The EPO case law had made it clear that computer-assisted methods are an accepted means for determining the practical utility of an invention. Moreover, if a certain degree of homology is shown using such methods, industrial applicability would be acknowledged if the other conditions for showing industrial application are also fulfilled.

In addition to this set of principles, it is significant that with these decisions, the EPO has not only answered questions regarding the interpretation of Rule 29(3) of the EPC as applicable to human DNA inventions, but it has further clarified the interpretation of the requirement of industrial application in general. These decisions have revived the main principles behind the criterion of industrial application, thus reemphasizing its important role within the patent system. In fact, most of the criteria that arise from the EPO case law on the industrial applicability of human gene inventions are now part of the interpretative context of Article 57 EPC that is applicable to all technology fields. In particular, as noted in Chapter 3, the requirement of showing a profitable use of an invention in industry in the sense of an immediate concrete benefit, as opposed to vague and speculative indications of possible objectives that might or might not be achievable by carrying out further research, are explicitly included in the EPO compilation of case law for interpreting industrial application in general.[168] This corresponds to the reality that gene patenting may decrease over time and inventions in other research areas that are rapidly evolving may pose similar challenges for the assessment of the industrial applicability of such inventions in the future.

With regard to the question of whether Rule 29(3) of the EPC has effectively imposed a barrier to the patenting of gene sequences of unknown function, there are now a significant number of decisions confirming that the requirement of industrial application has raised the bar for the patentability of human genes. Some commentators sustain that the new approach has raised the industrial applicability requirement from a non-existent standard to an unacceptably high standard,[169] and that Rule 29(3) of the EPC now provides major obstacles to the granting of claims for ESTs, especially when they are to be used as probes for the isolation of full-length genes of unknown function.[170] They argue that the test of industrial application is now of a higher standard in cases involving gene sequences than for other inventions,[171] which also raises doubts regarding the compatibility of Rule 29(3) with Article 27 of the TRIPS Agreement. In particular, the difference between the European industrial applicability requirement that is applicable to human DNA inventions and inventions in other technical fields suggests that the non-discrimination principle, as stated by the TRIPS Agreement, has no direct effect.

However, the pragmatic approach that the EPO has taken towards interpreting both Article 57 and Rule 29(3) of the EPC within the context of human gene patents shows that the current industrial application criteria meet the challenges posed by gene technologies by providing a suitable barrier to the grant of patents on inventions of a speculative nature. Yet it provides flexibilities in those cases where an invention's

---

[168] Case law of the Boards of Appeal of the European Patent Office, Seventh Edition (September 2013) section I.E.1.2: Indication of a profitable use of the invention in industry.

[169] Joshua C Benson, 'Resuscitating the Patent Utility Requirement, Again: A Return to *Brenner v. Manson*' (2002) 36 UC Davis Law Review 267.

[170] Aerts, 'The Industrial Applicability and Utility Requirements for the Patenting of Genomic Inventions: A Comparison between European and US Law' (n 33).

[171] Oliver Mills, *Biotechnological Inventions, Moral Restraints & Patent Law* (Revised Edition, Ashgate 2010) 142.

function is known to be especially important for human health. Nonetheless, given the importance of striking a balance between public and industrial interests, it would be desirable for the EPO to keep an impartial, realistic and pragmatic implementation of the industrial application criteria for these types of inventions.

Finally, it should also be noted that the nature of the activities that are covered under the notion of 'profitable use' is bound to evolve. While a disclosure may not generate a profitable use in the sense currently accepted, it may still be of value for some of the intermediate activities in which companies are now increasingly involved.[172] Thus there is the possibility that this test may be further broadened in the future to cover new activities in relation to gene technologies.

4.5 THE CASE OF NEUTROKINE ALPHA: *ELI LILLY v HUMAN GENOME SCIENCES*

The 2009 EPO decision and the 2011 decision of the UK Supreme Court in the case of *Eli Lilly v HGS* have provided additional conclusions with regard to the interpretation of the requirement of industrial applicability for human gene inventions. This case illustrates very well the EPO's position on the interpretation of Rule 29(3) of the EPC and provides a good example of how the EPO Boards apply previous EPO decisions on industrial application consistently. Furthermore, this case shows the differences between the approaches of the UK courts and the EPO towards the industrial applicability of isolated human proteins and their encoding genes, and how coherence between their decisions has eventually been achieved with regard to this question.

In 1996, the US biotech company HGS, using homology screening, identified a new member of the tumour necrosis factor (TNF) ligand superfamily, which they named Neutrokine alpha (Neutrokine-α). Subsequently, HGS filed a patent application at the EPO that included claims to the nucleic and amino acid sequences of Neutrokine-α, therapeutic and diagnostic applications of the protein, and antibodies to the protein.[173] The application described an extensive, but slightly contradictory, list of potential activities and uses of Neutrokine-α based on its membership of the TNF ligand superfamily.[174] These included:

---

[172] Leverve (n 54).

[173] European Patent EP-B-0939804, Neutrokine alpha (1996).

[174] As Kitchin J said, the specification 'contains extravagant and sometimes contradictory claims', see *Eli Lilly & Co v Human Genome Sciences Inc* [2008]

- activities that are believed to be useful for both immune suppression and enhancement
- effects on the mobilization, proliferation, differentiation and migration of stem cells and
- the treatment of leukaemia, among others.

However, the patent application as filed did not provide any experimental data that could demonstrate the claimed biological properties and activities of Neutrokine-α.

Eli Lilly & Company was involved in a similar research project and had also independently identified Neutrokine-α sometime after HGS. Following the grant of HGS's patent in August 2005, Eli Lilly initiated opposition proceedings before the EPO and revocation proceedings before the English High Court, where the applicant had designated the patent.[175] Eli Lilly argued with regard to industrial application that the HGS's invention was not capable of practical utility. The company alleged that the application was speculative and that it did not disclose the invention clearly and completely enough for it to be performed by a skilled person in the art.

Before the High Court gave its judgment, the patent had already been found invalid in the EPO opposition proceedings on grounds of added matter (Article 123(2) EPC) and lack of inventive step (Article 56 EPC). In contrast, the High Court finally held the patent invalid for lack of industrial application, insufficiency and obviousness. However, although the EPO held the patent invalid on a different ground, the EPO OD and the High Court adopted similar understandings in reaching their decisions. With regard to industrial application, the High Court noted in its reasoning for the decision that there were only two UK decisions regarding industrial application at that time.[176] For this reason, Kitchin J decided to carry out an exhaustive review of the EPO case law on this

EWHC 1903 (Pat), [2008] RPC 29, para 134. Perhaps rather more tolerantly, the TBA at the EPO referred to the patent as having been drafted on a 'boiler-plate' basis, which it described as 'a practice used by patentees', see *Neutrokine alpha/HUMAN GENOME SCIENCES* (T 0018/09) [2009] (EPO (TBA)) point 27.

[175] *Eli Lilly & Co* (EWHC) (n 174).

[176] *Chiron Corporation* (n 93) where the English Court of Appeal held that the claims in the patent to polypeptides which did not have any known use or purpose, for example encoding the Hepatitis C virus, at the time of filling the patent application were invalid for lack of industrial application; UKIPO patent BL O/286/05, *Aeomica Inc* (2005); *Eli Lilly & Co* (EWHC) (n 174) para 186.

subject,[177] in addition to considering the 'specific substantial and credible' test introduced by the USPTO in *Brenner v Manson*.[178] Following this, Kitchin J formulated a set of nine principles for assessing the utility of genetic inventions.[179] These principles, by reflecting the EPO approach towards the interpretation of the industrial applicability requirement, posed a higher barrier for the patentability of inventions concerning human genes and related proteins.

Subsequently, parallel appeal proceedings were heard before the EPO and the English Court of Appeal. The Court of Appeal decided to stay proceedings until the EPO had reached its decision; however, the procedural cooperation between tribunals finally resulted in conflicting findings. The TBA at the EPO upheld the European patent on the basis of more restricted claims. It found that the application disclosed a concrete technical basis for the skilled person to recognize an industrial exploitation of the claimed invention.[180] Nevertheless, Jacob LJ dismissed the appeal at the English court for lack of some reasonable, credible use for any member of the TNF ligand superfamily.[181] Jacob LJ attempted to explain why the outcome of the court had differed from the EPO decision based on the differences between the decision-making processes in the English courts and the EPO. English trials at first instance, which involve intensive investigation and testing of evidence, were contrasted with the opposition procedure at the EPO, which was referred to as 'administrative' because it usually accepts the admission of fresh material on appeal and does not include cross-examination or compulsory disclosure

---

[177] *Icos Corporation* (n 115); *Multimeric Receptors* (n 132); *Max Plank* (n 135); *Zymogenetics* (n 117); *Bayer* (n 145); *IL-17 related polypeptide/ SCHERING* (T 1165/06) [2007] (EPO (TBA)).

[178] *Brenner v Manson* (n 61).

[179] The nine principles identified by Kitchin J can be summarized as follows: (i) the notion of industry must be construed broadly; (ii) the industrial application must be identified by the person skilled in the art; (iii) the description must disclose a practical way of exploiting the invention; (iv) a real prospect of a concrete application in industry must be directly derivable from the specification; (v) research results that require further research are not enough; (vi) patents should not grant unjustified control over an unexplored field of research; (vii) use for human health purposes is enough; (viii) the claimed invention must not be used to find out more about its activities; (ix) using bioinformatics techniques to identify the function of the claimed substance is allowed. See *Eli Lilly & Co* (EWHC) (n 174) para 226.

[180] *Neutrokine alpha* (n 174).

[181] *Eli Lilly & Co v Human Genome Sciences Inc* [2010] EWCA Civ 33, [2010] RPC 14.

documents. Thus, the evidence before the EPO and the English Court of Appeal and the manner in which they considered and tested it were remarkably different. Therefore, although the EPO and the English Court of Appeal were applying the same principles, the standards by which those principles were applied differed between them. Furthermore, Jacob LJ reiterated that although the UK courts should follow any principle of law that is laid down by the TBA, they still reserve the right to differ if this implies that the commodore would 'steer the fleet on to the rocks'.[182]

Not satisfied with the Court of Appeal's finding, HGS brought the case before the UK Supreme Court,[183] which in November 2011 handed down its first judgment on a patents case. Lord Neuberger overturned the earlier decisions of the Court of Appeal and the High Court and brought the UK position on industrial application into line with the EPO approach. In doing so, Lord Neuberger first reviewed Article 5 of the Biotech Directive, domestic and US jurisprudence, and the case law of the EPO.[184] He then referred to submissions by the Bioscience Industry Association (BIA), arguing that raising the bar on industrial application would cause major damage to the biotech industry. Following this, Lord Neuberger elucidated fifteen principles from the TBA jurisprudence that were applicable to Article 57.[185] In view of these principles, which essentially overlap with the nine principles described by the High Court,

---

[182] *Actavis v Merck* [2008] EWCA Civ 444, [2009] 1 WLR 1186.

[183] *Eli Lilly & Co v Human Genome Sciences Inc* [2011] UKSC 51, [2012] RPC 6.

[184] Point 2 of the UKSC decision also remarks on the case-by-case basis upon which the requirement of industrial application is assessed. Furthermore, Lord Neuberger acknowledged the importance of biotech research and the need for caution in order to avoid potentially far-reaching consequences for scientific research, the biotech industry and human health.

[185] The general principles identified by the UK Supreme Court embrace the nine principles described by the High Court and add the following ideas. When the patent discloses a new protein and its encoding gene, merely identifying the structure of a protein, without attributing to it any 'practical use' is not enough; however a 'plausible' or 'reasonably credible' claimed use, or an 'educated guess', can suffice and can be assisted by being confirmed by later evidence; and the requirements of a plausible and specific possibility of exploitation can be at the biochemical, cellular or biological level. When the protein is said to be a family or superfamily member, assigning a similar role to the protein is enough since the problem to be solved can be 'isolating a further member of the family'; however this is not enough if the claimed role of the protein or its membership of the family is questionable, or if the known members of the family have different functions. See *Eli Lilly & Co v Human Genome Sciences Inc* (UKSC) (n 183) para 107.

Lord Neuberger found that the information disclosed in the patent application on Neutrokine-α was sufficient to satisfy the requirement of industrial application.

Several important conclusions can be drawn from the final resolution of this case. First, this case provides a very good example of how national courts and the EPO, despite applying the same standard, can arrive at different conclusions due to differences in the decision-making process and the kind and amount of evidence that is examined. In this case, both the English Court of Appeal and the TBA at the EPO were relying on the same EPO decisions and criteria for assessing the claimed invention's compliance with Rule 29(3) of the EPC. However, as Jacob LJ himself indicated, the English courts' more rigorous procedure with regard to the investigation and testing of evidence, as opposed to the EPO's more relaxed requirements regarding the provision of evidence, may lead to a different outcome.

Second, this case shows that although the national courts of EPO Member States should follow the EPO's interpretation of the EPC, they may recognize the EPO's approach but still decide not to apply the EPC provisions in the same way and to develop their own interpretative criteria. Nonetheless, in its decision the UK Supreme Court expressed its intention to harmonize the UK's interpretation of the EPC with that of the EPO and other contracting States.[186] This express declaration could be seen as a significant step towards consistency between the decisions of the EPO and national courts. In this regard, the fact that the final decision of the UK Supreme Court followed the approach taken by the EPO TBA further confirms this intention and may encourage national courts to work on reducing differences with the EPO and other European countries in this or other patent law areas.

Finally, the finding of the UK Supreme Court shows judicial concerns about the antagonistic forces faced by inventors when deciding whether to file a patent application. On the one hand, if an inventor files a patent application early without sufficient evidential data to support their claims, they risk both not obtaining the patent and having publically disclosed their invention. However, if the inventor waits until further experimental data are obtained, there is the risk that a competitor will obtain the patent first. In this regard, the UK Supreme Court relied on the policy argument concerning the need to reduce the risk of discouraging investment in bioscience, whereby a too high threshold to satisfy the requirement of industrial application may discourage innovators from investing in

---

[186] *Eli Lilly & Co v Human Genome Sciences Inc* (UKSC) (n 183) para 171.

research and development. Nonetheless, although the UK Supreme Court's acknowledgement of such strong industrial interests in its decision may be seen as indicative of the courts' relaxation towards the assessment of industrial applicability, no other decisions from the EPO or other European courts indicate that this is the case.

4.6 CONCLUDING REMARKS

Before the adoption of the Biotech Directive there were few decisions interpreting the requirement of industrial application and the EPO and national patent offices often relied on extraordinary examples to illustrate when an invention was not capable of industrial application. However, with the introduction of EPC Rule 29(3) the EPO case law on the industrial application of human DNA sequences has progressively provided a mechanism to raise the bar of the patentability of genetic inventions of speculative function. Furthermore, the EPO decisions on the industrial application of human gene sequences have not only answered key questions with regard to the interpretation of the standard set by Article 5(3) of the Biotech Directive, but they have also clarified important issues for the interpretation of the general requirement of industrial application as applicable to all fields of technology. These decisions have revived the fundamental ideas behind the criterion of industrial application, thus reemphasising its important role within the patent system.

This chapter aimed to offer a detailed analysis of the background, meaning and practical implications of the requirement introduced by Article 5(3) of the Biotech Directive (later incorporated into Rule 29(3) of the EPC) of disclosing the industrial application of sequences and partial sequences of human genes in the patent application. It provides the basis for understanding the following chapters, which will focus on the implications of Article 5(3) of the Biotech Directive in the interpretation of the exclusion of discoveries with regard to human DNA sequences and in determining the scope of protection for these types of inventions.

5. The requirement of industrial application and the interpretation of the exclusion of human genetic discoveries from patent protection

5.1 INTRODUCTION

The distinction between discoveries and inventions is especially difficult in cases where the claimed subject matter is a newly discovered genetic sequence rather than a final product incorporating it. The EPO has dealt with this issue by implementing the so-called 'isolation principle', according to which the process of separating the genetic material from its natural source can give a discovered substance the character of invention. However, after the adoption of the Biotech Directive, 'isolation' is no longer enough for a discovered substance to be patented, at least for inventions concerning human genes. Article 5(3) of the Biotech Directive introduced the additional requirement of disclosing the industrial application of isolated human DNA sequences for such substances to acquire the status of invention. Therefore, since the Biotech Directive entered into force, in order to become an invention the claimed human genetic material needs to be isolated and its utility also has to be disclosed in the patent application. Thus, studying the role of the requirement of industrial application in 'transforming' isolated newly discovered human genetic substances into patentable inventions is essential for understanding the relationship between the exclusion from patentability of abstract discoveries and the requirement of industrial application, as well as the interpretation of the notion of invention within the context of human gene patents.

To provide such understanding, this chapter first studies the exclusion from patentability of discoveries and the inherent requirement for an invention under the EPC, and the role of the requirement of practical applicability in the assessment of the requirement for an invention in patent law. The chapter then examines the exclusion from patentability of discoveries concerning human genetic material to then explore the EPO isolation approach towards the patentability of naturally occurring substances. Finally, the chapter analyses the Biotech Directive's introduction of Article 5(3) as a new element in the distinction between discoveries and inventions in the case of human genes. It discusses the justification, interpretation and implications of such approach.

5.2 THE EXCLUSION FROM PATENTABILITY OF DISCOVERIES IN EUROPEAN PATENT LAW

5.2.1 Discoveries as Excluded Subject Matter under the EPC

A basic premise of European patent law is that patents are to be granted for inventions as opposed to abstract subject matter such as discoveries. A central problem of the patent system is therefore to distinguish what is patentable from what is not, taking the notion of invention as the reference standard.[1] Article 52(1) of the EPC (which corresponds with Article 27(1) of the TRIPS Agreement) makes clear that only inventions can be protected by patent rights. However, neither the TRIPS Agreement nor the EPC provide a definition of 'invention' in patent law terms.

In respect of patent rights, the TRIPS Agreement aims inter alia to reduce differences between the patent systems of the World Trade Organization (WTO) Member States and achieve technological neutrality and effective enforcement of patent law.[2] It requires Member States to provide equal protection without discrimination for inventions in all fields of technology.[3] However, the text of the TRIPS Agreement gives certain flexibility to Member States to define for themselves what an invention is, as well as for excluding specific types of subject matter from patent protection, for instance, diagnostic methods for the treatment of humans, plants, and animals other than microorganisms, or inventions the commercial exploitation of which is necessary to protect public order or morality, as well as to protect human or animal health.[4] Nonetheless, the absence of an express definition of invention does not imply that

---

[1] See Edmund W Kitch, 'The Nature and Function of the Patent System' (1977) 20 Journal of Law & Economics 265.

[2] Agreement on Trade-Related Aspects of Intellectual Property Rights of 15 April 1994, Preamble.

[3] Ibid, art 27(1).

[4] Ibid, art 27(2) and (3). Thus, in spite of the overall goal of the agreement relating to fostering harmonization of patent law, the TRIPS Agreement allows some differences in the patentability of certain technologies between Member

countries can domestically expand the scope of patentable subject matter without limitation.[5]

In Europe, the patent law notion of invention is defined by the requirements and exclusions of Article 52 of the EPC. The requirement for an invention is set in Article 52(1) in conjunction with Articles 52(2) and 52(3) of the EPC, the latter two referring to a number of exclusions that are not considered inventions and thus are excluded from patent protection.[6] What the exclusions under Article 52(2) (a) to (d) have in common is that they refer to activities that do not aim for any direct technical result but are rather of an abstract and intellectual character,[7] and thus do not fit in with the objectives of the patent system of promoting the creation of innovative technical products and processes rather than abstractions.[8] Among others, Article 52 of the EPC excludes discoveries 'as such'[9] from patent protection.

The difference between discoveries and inventions is an elementary feature of virtually all patent law regimes,[10] especially since the exclusion of discoveries and the requirement for an invention have particular meanings in patent law, which reflect fundamental policy objectives including the encouragement of innovation and the reward of endeavour.[11] Therefore, drawing an appropriate boundary between unpatentable, merely discovered phenomena and patentable inventions is essential to prevent the patent system from unduly restricting access to fundamental

---

States. See also Nuno Pires de Carvalho, *The TRIPS Regime of Patent Rights* (Kluwer Law International 2002) 145.

[5] See Luigi Palombi, 'The Impact of TRIPS on the Validity of the European Biotechnology Directive' [2005] JIBL 62.

[6] Justine Pila, 'On the European Requirement for an Invention' (2010) 41 IIC 906.

[7] Case law of the Boards of Appeal of the European Patent Office, Seventh Edition (September 2013) I-A.2.2: 'Discoveries, scientific theories and mathematical methods'. See also *Multimeric Receptors/SALK INSTITUTE* (T338/00) [2002] (EPO (TBA)) point 3, dealing specifically with the exclusion of discoveries under Article 52(2)(a) of the EPC; Gerald Kamstra and others, *Patents on Biotechnological Inventions: The EC Directive* (Sweet & Maxwell 2003) 10.

[8] Peter Drahos, 'Biotechnology Patents, Markets and Morality' (1999) 21 EIPR 441.

[9] Convention on the Grant of European Patents (European Patent Convention (EPC)) of 5 October 1973, art 52(2)(a) and (3).

[10] Matthias Herdegen, 'Patenting Human Genes and Other Parts of the Human Body under EC Biotechnology Directive' (2001) 4 BIO-Science Law Review 102.

[11] Hector L MacQueen, *Contemporary Intellectual Property: Law and Policy* (2nd edn, Oxford University Press 2011) 512.

scientific discoveries[12] that do not reach the standards of creativity and technicality necessary to be granted patent protection.

In non-legal terms, a discovery is the act of uncovering something for the first time. It has been frequently described as finding something that was previously unknown but was pre-existing in nature, or as the identification of properties and utilities unknown so far, and thus not part of the legal definition of state of the art.[13] The Geneva Treaty on the International Recording of Scientific Discoveries of 3 March 1978 defines discovery as 'the recognition of phenomena, properties or laws of the material universe not hitherto recognized and capable of verification'.[14] It therefore follows that a discovery is not the result of creation efforts, even if creativity was needed to reveal information concealed in nature;[15] but its merit resides in a different type of work, directed at uncovering a reality rather than creating it, or on the discoverer's serendipity.

Thus, they fall into the category of concepts or knowledge that already exist and are just waiting to be discovered by man.[16] Moreover, because of their abstract character, discoveries are consequently not yet directly applicable in industry and not of obvious industrial application.[17] Both inventions and discoveries contribute to the advance of knowledge, but inventions do it through the application of that knowledge,[18] while discoveries need to be further developed to become applicable.[19]

---

[12] Christopher M Holman, 'Patent Border Wars: Defining the Boundary between Scientific Discoveries and Patentable Inventions' (2007) 25 TRENDS in Biotechnology 539. See also Sigrid Sterckx and Julian Cockbain, *Exclusions from Patentability: How Far Has the European Patent Office Eroded Boundaries?* (Cambridge University Press 2012) 129.

[13] Giuseppe Sena, 'Directive on Biotechnological Inventions: Patentability of Discoveries' (1999) 30 IIC 731; Li Westerlund, *Biotech Patents: Equivalence and Exclusions under European and U.S. Patent Law* (Kluwer Law International 2002) 24.

[14] Geneva Treaty on the International Recording of Scientific Discoveries of 3 March 1978, art 1(1)(i).

[15] Pires de Carvalho, *The TRIPS Regime of Patent Rights* (2002) (n 4) 146.

[16] Westerlund (n 13) 24 and 49.

[17] Friedrich-Karl Beier, R Stephen Crespi and Joseph Straus, *Biotechnology and Patent Protection: An International Review* (OECD Publishing 1985) 55.

[18] MacQueen (n 11) 512. Another related important distinction is that between the invention (idea), and the embodiment (the material artefact in which it is expressed), see Alain Pottage and Brad Sherman, *Figures of Invention: A History of Modern Patent Law* (Oxford University Press 2010) 1 and 22.

[19] Pires de Carvalho, *The TRIPS Regime of Patent Rights* (2002) (n 4) 146–147.

Although discovered knowledge can be new in the sense that nobody knew about its existence, this knowledge per se will have difficulty finding a concrete application in industry. This fact makes the patent law requirement of industrial application a key factor in the distinction between inventions and excluded discoveries.

The EPC does not define discoveries, but only excludes them from patentability due to their abstract character or lack of technical character.[20] Moreover, neither the EPC nor the EPO practice explain what 'abstract' means in patent law terms. With regard to the exclusion of discoveries, the EPO Guidelines provide that if a new property of a known material or article is found, it would constitute a mere discovery and would be unpatentable; but if, however, such new property can be put to practical use, the result is an invention of technical character that may be patentable.[21] Thus, the line of demarcation between the mere discovery of a natural substance and its characterization as an invention would depend on the degree of human technical intervention that has been necessary to develop that discovery into something that has a practical use.[22] It seems that it is in the explanation of what gives an invention technical character that, although in negative terms, the definition of discovery as such in terms of patent law may be found.

On the other hand, pursuant to Article 52(3) of the EPC, the exclusion of discoveries only refers to discoveries 'as such'. The inclusion of this limitation shows the EPC's aim of narrowing the scope of exclusions as much as possible.[23] In practice, the 'as such' approach has led to an 'anything added' approach that allows the patentability of discoveries as

---

[20] Guidelines for Examination in the European Patent Office, as last revised in November 2014, Part G-II, 3.1.

[21] Ibid.

[22] Commission of the European Communities proposal for a Council Directive on the legal protection of biotechnological inventions (1988) [COM(88) 496 final – SYN 159] OJ C10/3, 40–41. This is consistent with the longstanding practice, even prior to the advent of modern biotechnology, of granting patents for isolated and purified chemical products that exist in nature only in an impure state, see Rebecca S Eisenberg, 'How Can You Patent Genes?' (2002) 2 The American Journal of Bioethics 3.

[23] EPC (n 9) art 52(3). See EPO Case Law of the Boards of Appeal (n 7) I.A.1.1, explaining that the intention of Article 52(3) of the EPC is to ensure that anything which had previously been a patentable invention under conventional patentability criteria should remain patentable under the EPC (*Estimating sales activity/DUNS LICENSING ASSOCIATES* (T 0154/04) [2006] (EPO (TBA)) point 7), and that it was the clear intention of the contracting states that this list of 'excluded' subject matter should not be given a too broad scope of application,

long as something else is added by man.[24] This interpretation of the 'as such' limitation has had a profound impact on the scope of the exclusion of discoveries from patentable inventions. Narrowing the scope of the exclusion of discoveries has in turn resulted in a corresponding expansion of the scope of subject matter eligible for patent protection.[25] Such expansion has been especially controversial in the field of biotechnology, mainly due to the presence of pre-existing natural elements that have been discovered by man in all inventions within this field.

In contrast, an invention can be defined, in non-patent law terms, as a new product or process with no previous existence. It is the creation of something that did not exist before[26] and that improves the objective world existing in nature.[27] Moreover, all inventions are the result of human intelligence. A phenomenon is a discovery rather than an invention if human intelligence has not intervened in obtaining it.[28] In this regard, a central question would be whether the completed invention depends on, or is a result of, the mental activities of the person who carries it out.[29] However, this does not mean that an invention cannot be based on a previous discovery.[30] In many occasions, discoveries can be

---

as follows from the legislative history of Article 52(2) EPC (*Estimating sales activity* (n 23) point 6).

[24] Sterckx and Cockbain (n 12) 117.

[25] See for example MacQueen (n 11) 509; Pila, 'On the European Requirement for an Invention' (n 6).

[26] Sena (n 13).

[27] Westerlund (n 13) 49. The assumption that inventions are new creations also raises philosophical questions. From a platonistic conception, all possible knowledge already exists and only needs to be discovered. Thus, humankind does not actually create anything, but only discovers it; see Platon's Theory of Forms discussed in Reinier Bakels, *The Technology Criterion in Patent Law: A Controversial but Indispensable Requirement* (Wolf Legal Publishers 2012) 180.

[28] Westerlund (n 13) 51.

[29] See Sterckx and Cockbain (n 12) 115.

[30] EPO Case Law of the Boards of Appeal (n 7) I-A.2.2.1, explaining that if a mere discovery is put to practical use, then it can constitute an invention which may be patentable. See also *Friction reducing additive* (G 0002/88) [1989] (EPO (EBA)) point 8 explaining that the fact that the idea or concept underlying the claimed subject-matter resides in a discovery does not necessarily mean that the claimed subject-matter is a discovery 'as such'; *Genentech Inc's Patent* [1989] RPC 147 (Civ) per Purchas LJ, para 12.09, explaining that not all discoveries come within the exclusion in Section 1(2)(a) of the 1977 Patents Act (which corresponds with Article 52(2)(a) of the EPC), and also stating that a discovery is capable of forming the substratum of invention if it is applied in a technique or process or incorporated in a product; EPO Examination Guidelines (n 20) Part

put to practical use and can give rise to important inventions. For example, since the discovery of epinephrine (adrenaline) in 1901, this hormone has been the basis of multiple inventions in the pharmaceutical field[31] that are very important in the treatment of many health problems such as cardiac arrest or asthma.

Moreover, often discoveries and inventions are identical from the point of view of the amount of work behind a concrete result. The research activity that leads to the identification of unknown, although pre-existing, phenomena or to the determination of their unknown characteristics substantially equates to the activity leading to the creation of a new product.[32] Therefore, it is not the level of effort behind discovering and inventing something that determines when there is a real invention. The principal question in the discovery-invention dichotomy is at what point along a line of research an addition to scientific knowledge adds enough technical character to transform a mere discovery into an invention.[33] In this regard, it is essential to determine what constitutes technical character within the context of patent law.

There is little case law on the specific exclusion from patentability of discoveries but the EPC deals with this issue in the EPC Examination Guidelines,[34] where special emphasis is given to the distinction between discoveries and inventions within the context of biological patents.

5.2.2 The Requirement for an Invention in the EPC

While the EPC clearly states that patent protection is only available for inventions, it does not include a precise definition of what an invention is in terms of patent law. It only provides an open list of subject matter that are not inventions and are thus excluded from patent protection, and a set of requirements that an invention has to comply with in order to be

---

G-II, 3.1: 'if, however, that property is put to practical use, then this constitutes an invention which may be patentable. For example, the discovery that a particular known material is able to withstand mechanical shock would not be patentable, but a railway sleeper made from that material could well be patentable'.

[31] Lori B Andrews, 'The Gene Patent Dilemma: Balancing Commercial Incentives with Health Needs' (2002) 2 Hous J Health L & Pol'y 65.

[32] Sena (n 13).

[33] See William Cornish, David Llewelyn and Tania Aplin, 'Intellectual Property in Biotechnology' in *Intellectual Property: Patents, Copyright, Trade Marks & Allied Rights* (7th edn, Sweet & Maxwell 2010) 924.

[34] EPO Examination Guidelines (n 20) Part G-II, 3.1.

patentable.[35] Apart from this, there are no other provisions within the EPC explaining what gives a phenomenon the character of invention. However, determining what is and what is not an invention is essential for understanding inherent patentability,[36] as well as for demarking a clear distinction between inventions and discoveries.

When considering the EPC and the Implementing Regulations all together, the concept of invention can be defined as a technical solution (pertaining to a technical field and having technical features/teaching or producing a technical effect) to a technical problem.[37] Presumably, the requirement for an invention to have technical character, or in other words, to provide a technical contribution to the art, is based on a long-standing legal practice in at least the majority of Contracting States of the EPO.[38] In this regard, several EPO decisions acknowledge that the need for technical character is an implicit, fundamental pre-requisite for all inventions.[39] It is a pre-patent determinant, or a condition sine qua non, for all inventions.

Thus a given phenomenon may be an invention within the meaning of Article 52(1) of the EPC if, for example, a technical effect is achieved by the invention or if technical considerations are required to create the

---

[35] EPC (n 9) art 52. See also *Estimating sales activity* (n 23) point 6: 'The application of Article 52(1) EPC presents a problem of construction as there was no legal or commonly accepted definition of the term "invention" at the time of conclusion of the Convention in 1973. Moreover, the EPO has not developed any such explicit definition ever since.'

[36] Justine Pila, 'The Future of the Requirement for an Invention: Inherent Patentability as a Pre- and Post-Patent Determinant' in Emanuela Arezzo and Gustavo Ghidini (eds), *Biotechnology and Software Patent Law* (Edward Elgar Publishing 2011) 56.

[37] EPC (n 9) art 52, rule 42(1)(a) and (c), and rule 43(1). Technical teaching is defined as an instruction addressed to a skilled person as to how to solve a particular technical problem using particular technical means, see EPO Case Law of the Boards of Appeal (n 7) I-A.1.1. See also Kamstra and others (n 7) 10; Pila, 'The Future of the Requirement for an Invention: Inherent Patentability as a Pre- and Post-Patent Determinant' (n 36) 57.

[38] *Document abstracting and retrieving/IBM* (T 0022/85) [1988] (EPO (TBA)) point 3.

[39] Ibid, points 2 and 4; *Computer program product/IBM* (T 1173/97) [1998] (EPO (TBA)) point 5.1; *Controlling pension benefits system/PBS PARTNERSHIP* (T 0931/95) [2000] (EPO (TBA)) point 2; *Auction method/HITACHI* (T 0258/03) [2004] (EPO (TBA) point 3.1; *Odour selection/QUEST INTERNATIONAL* (T 0619/02) [2006] (EPO (TBA)) point 2.2.

invention.[40] Patented inventions can therefore be defined as technically inventive, new and useful solutions to technical problems, while discoveries fail to meet the requirement of technical character necessary to be an invention, and thus fail to meet the main criterion of patentability under Article 52(1).

In this regard, in Friction reducing additive, the EPO Enlarged Board of Appeal pointed out that the essential difference between inventions and discoveries is the lack of existence of technical character. This case concerned an invention claiming a new use for a known compound for a particular purpose, based on a technical effect described in the patent. When giving its decision, the board stated that:[41]

> In particular cases it may clearly be necessary to consider and decide whether a claimed invention is a discovery within the meaning of Article 52(2)(a) EPC. An essential first step in such consideration is ... to determine its technical features. If, after such determination, it is clear that the claimed invention relates to a discovery or other excluded subject-matter 'as such' ... then the exclusion of Article 52(2) EPC applies.

Even so, the EPC does not define 'technical character'. Although it is clear that abstract phenomena such as discoveries do not possess technical character, there is no actual basis for this requirement in the EPC.[42] Rules 42 and 43 of the EPC Implementing Regulations only state that the invention must relate to a technical field and solve a technical problem, and that the claims shall define the invention in terms of its technical features.[43] Thus, apart from the clarification of what types of subject matters lack technical character, the EPC does not provide any substantially sound legal explanation of what constitutes technical character to rely on.

Since the term 'technical' is not specifically defined, it has to be interpreted on a case-by-case basis. Nonetheless, a few keynotes can be

---

[40] EPO Case Law of the Boards of Appeal (n 7) I-A.1.1. See also EPO Examination Guidelines (n 20) Part G-II, 3.1.

[41] *Friction reducing additive* (n 30) point 8. See also EPO Examination Guidelines (n 20) Part G-II, 3.1.

[42] In fact, the word 'technical' appears only in the context of the qualifications (legal or technical) of members of the Boards of Appeal. See Philip W Grubb and Peter R Thomsen, *Patents for Chemicals, Pharmaceuticals, and Biotechnology: Fundamentals of Global Law, Practice, and Strategy* (Oxford University Press 2010) 70.

[43] EPC (n 9) rules 42 and 43. See also EPO Examination Guidelines (n 20) Part G-I, 2.

identified. First, it can be noted that the notion of the term 'technical' in patent law is very broad and has no clear boundaries as to its precise meaning. Under the EPO system, patents are available for inventions in all fields of technology, so the requirement for an invention to belong to a technical field is interpreted very broadly and includes unconventional fields such as agriculture.[44] Moreover, with the rapid development of modern technologies such as genetic engineering, the tendency has been to expand the number of fields for which patent protection is available.

Second, technical character does not mean that natural elements cannot be part of an invention. The concept of invention as used in patent law can be defined as a technical solution that utilizes laws of nature or controllable natural forces that cannot as such be obtained without the interference of human intelligence.[45] All human creations are ultimately based on natural sources. Thus, the presence of natural elements does not prevent an invention from having technical character. Such presumption is indeed especially relevant in the distinction between discoveries and inventions in the biotechnology field.

Third, despite the lack of a clear definition, it is presumed that being technical implies a relationship with applied knowledge or practical sciences as opposed to theoretical or abstract knowledge. The fact that inventions solve technical problems suggests that the industry is ready to receive and evaluate them,[46] something that would not happen in the case of abstract discoveries. The foundation of excluding discoveries from patentability is that they belong to the world of theoretical knowledge and do not have any form of technical utility.[47] In this sense, technical character has been described as requiring a human action on the physical world producing an objectively discernible material result directed at improving industrial activities, as distinct from other activities such as administrative or professional arts.[48] This notion of technical, as a condition related to the practical sciences and industrial activities reveals an important connection between the technicality of an invention and its applicability in industry. In fact the overlap of the invention requirement and the industrial application criterion is more pronounced in the case of gene patents (see section 5.4), although a close relationship between both

---

[44] EPC (n 9) arts 52(1) and 57.

[45] Westerlund (n 13) 38 and 51.

[46] Pires de Carvalho, *The TRIPS Regime of Patent Rights* (2002) (n 4) 146.

[47] R Stephen Crespi, *Patenting in the Biological Sciences: A Practical Guide for Research Scientists in Biotechnology and the Pharmaceutical and Agrochemical Industries* (Wiley 1982) 100.

[48] Pila, 'On the European Requirement for an Invention' (n 6).

concepts has also been highlighted in various European decisions concerning other kinds of inventions.[49] However, apparently, these two pre-conditions for patentability are not exactly the same. From a patent law perspective, industrial, practical and technical conceptions of patentable subject matter reflect somewhat different views on what should be the reference point for assessing which types of matter should be protected by patent law.

The requirement of industrial application plays a very important role in the distinction between discoveries and inventions since mere discoveries of an abstract character are unlikely to have any applicability in an industrial field.[50] This criterion essentially overlaps with the requisite of technical character since those subject matters that are technical are usually useful and have some applicability in industry. However, not all knowledge of industrial application is technical in the sense that it is understood by the EPO.[51] Although both requirements are intimately related,[52] they are not substitutes or synonymous, but are two distinct standards that have to be fulfilled independently.[53] Nonetheless, changes to the interpretation of industrial application with regard to human DNA inventions suggest that there is not such independence between the criteria of technical character and industrial applicability in the particular field of human genetics.

Moreover, given that the EPO does not define what technical means, assuming that something that has a practical function that can be used in

---

[49] Sivaramjani Thambisetty, 'Legal Transplants in Patent Law: Why Utility is the New Industrial Applicability' Law, Society and Economy Working Papers 6/2008 (2009) 49 Jurimetrics J 155.

[50] Westerlund (n 13) 41–42.

[51] *Odour selection* (n 39) point 2.6.2; *Undeliverable mail/PITNEY BOWES* (T 0388/04) [2006] (EPO (TBA)) points 4 and 6. EPO Examination Guidelines (n 20) Part G-III, 3: 'It should be noted that "susceptibility of industrial application" is not a requirement that overrides the restriction of Art. 52(2), e.g. an administrative method of stock control is not patentable, having regard to Art. 52(2)(c), even though it could be applied to the factory store-room for spare parts. On the other hand, although an invention must be "susceptible to industrial application" and the description must indicate, where this is not apparent, the way in which the invention is thus susceptible (see F-II, 4.9), the claims need not necessarily be restricted to the industrial application(s)'.

[52] *Card Reader/IBM* (T 0854/90) [1992] (EPO (TBA)) point 2.1. See also Indre Kelmelyte, 'Can Living Things Be Objects of Patents?' (2005) 2 International Journal of Baltic Law 1.

[53] *A method of functional analysis* (T 0953/94) [1996] (EPO (TBA)) point 3.11.

some kind of industry may not have technical character would impose a rather artificial interpretation of the notion of technicality. Technical applications essentially relate to practical applications of more abstract phenomena. Indeed, the EPO has constantly explained in its Guidelines for Examination that if a discovery is put into practice, it becomes an invention and can therefore be patentable if the other patentability requirements are met.[54] Therefore, saying that something that is applicable in industry, and thus has a practical use, is not necessarily technical in terms of patent law seems to be a rather artificial construction that does not provide any clarity but increases the level of uncertainty surrounding the European requirement for a technical invention. Nevertheless, even when considering both criteria as independent, they are still intimately related since being industrially applicable is essential for proving the existence of some practical and technical effect.

Finally, under the EPC system, there is an obligation to ensure that the subject matter of a patent application is scrutinized and not presumed. Article 27(1) of the TRIPS Agreement as well as Article 52(1) of the EPC expresses society's decision that the patent system is not fit for creations that are not inventions.[55] Not every claim that meets the conditions of patentability is a patentable invention; there must first be an invention.[56] It is important to make a clear distinction between the concept of invention as a general and absolute requirement of patentability and the relative criteria of novelty, inventive step and the requirement of industrial applicability.[57] Thus, in principle, the assessment of whether there is an invention should be strictly separated from and not confused with the examination of the other three patentability requirements.

In this regard, there are two essential questions to which all patent applications need to respond. First, it is necessary to analyse whether there is an invention in terms of patent law, and second, whether such invention fulfils the essential patentability criteria of novelty, inventiveness and industrial applicability. In other words, patentability under the European system involves satisfying a two-step test. Step one consists of analysing whether the claimed subject matter is an invention and, if so, proceeds to step two in order to determine whether the invention is new,

---

[54] EPO Examination Guidelines (1987), Part C-IV, 2.3.

[55] Nuno Pires de Carvalho, *The TRIPS Regime of Patent Rights* (3rd edn, Kluwer Law International 2010) 252.

[56] Palombi, 'The Impact of TRIPS on the Validity of the European Biotechnology Directive' (n 5).

[57] *Estimating sales activity* (n 23) point 5; EPO Examination Guidelines (n 20) Part G-II, 2.

inventive and industrially applicable.[58] There must be an invention, and if there is an invention, it must satisfy the requirements of novelty, inventive step and industrial applicability in order to be patentable. However, in the case of human gene patents, the requirements to be a technical invention and have an industrial (practical) application are very much intertwined.

5.3 THE EXCLUSION OF DISCOVERIES WITHIN THE CONTEXT OF BIOLOGICAL INVENTIONS

5.3.1 Inventions Concerning Naturally Occurring Substances: A Challenge to the Traditional Definition of Invention

The European patent law exclusion of discoveries from patent protection is especially relevant in the patenting of inventions concerning living material.[59] In particular, Article 52(2)(a) of the EPC raises significant questions when dealing with inventions over genetic material previously existing in the human body, like applications claiming patents on complete genes, DNA sequences, cDNA, ESTs or SNPs.[60] For example with regard to natural genes, their variants and their correlation with disease states, the identification of a pre-existing gene and its variants would represent a discovery in the non-legal sense of the term.[61] However, such discoveries may constitute inventions in patent law terms if a technical character is acquired, like applying in practice a discovery that had not previously been in effect.[62] In other words, the application of the exclusion of discoveries to pre-existing naturally occurring substances such as genetic sequences basically refers to the question of whether biological material can be considered to be an invention even if the characteristics of such material are identical to its in situ natural status.

The discovery of naturally occurring substances in patent law refers to the idea that natural substances that are the result of Mother Nature's

---

[58] Luigi Palombi, 'Patentable Subject Matter, TRIPS and the European Biotechnology Directive: Australia and Patenting Human Genes' (2003) 26 UNSWLJ 782.

[59] See First Directive Proposal (n 22) point 13.

[60] Timothy Caulfield, E Richard Gold and Mildred K Cho, 'Patenting Human Genetic Material: Refocusing the Debate' (2000) 1 Nature Reviews Genetics 227.

[61] See Sterckx and Cockbain (n 12) 115.

[62] Ibid 115.

handiwork are not inventions and therefore not patentable.[63] This doctrine is not unique to European patent law but is inherent in patent law in general.[64] Historically, patent examiners, courts and legislators have sought to define a clear distinction between inventions, which may be patentable, and mere discoveries of natural products, which are in principle not patentable. The rationale behind this is that the simple discovery of nature's creations is not an invention and must be accessible to everyone to utilize,[65] since nobody has added any input to such products but has only uncovered their existence.

Despite the apparently clear classification of natural substances within the category of discoveries, this doctrine has not been substantially defined, which has generated some confusion among both practitioners and their clients.[66] Further, previous understandings of property and life have changed following the impact of key developments in the biotech sciences that view living materials with inheritable characteristics as potential subjects of exclusive rights.[67] Moreover, it adds to the confusion that all tangible items, especially in biotechnology, are products of nature in the sense that nature provides the source materials. In this regard, the fact that something is found in nature may be viewed as a weak objection to patentability.[68] Furthermore, nature also invents practical solutions for technical problems, like in the case of mutations responding to hostile environments.[69] So the question is how to distinguish an invention from a

---

63 Naturally occurring substances can raise different patent law objections like lack of novelty or inventiveness, but the main challenge is the exclusion of discoveries as patentable subject matter, see John M Conley and Roberte Makowski, 'Back to the Future: Rethinking the Product of Nature Doctrine as a Barrier to Biotechnology Patents (Part I)' (2003) 85 JPTOS 301.

64 Trevor Cook, *A User's Guide to Patents* (A&C Black 2011) 386. The EPC exclusion of discoveries of naturally occurring substances essentially corresponds with the US patent law exclusion of products of nature. For a discussion about the US case law on the 'product of nature' doctrine, see Conley and Makowski (n 63); Kelmelyte (n 52); Sterckx and Cockbain (n 12) 130–134.

65 John M Conley, 'Gene Patents and the Product of Nature Doctrine' (2009) 84 Chicago-Kent L Rev 109.

66 Karl Bozicevic, 'Distinguishing Products of Nature from Products Derived from Nature' (1987) 69 JPTOS 415.

67 Westerlund (n 13) 35. See also Warren D Woessner, 'The Evolution of Patents on Life – Transgenic Animals, Clones and Stem Cells' (2001) 83 JPTOS 830.

68 M Jacob, 'Patentability of Natural Products' (1970) 52 J Pat Off Soc'y 473.

69 Pires de Carvalho, *The TRIPS Regime of Patent Rights* (2010) (n 55) 253.

product of nature for the purposes of patent law.[70] Yet it appears that the answer to this problem is more a policy issue than a legal one.

In the case of human DNA inventions, the exclusion of discoveries has always played a relatively large role in comparison with other technical fields. For inventors in the field of genetics, the exclusion of discoveries poses the clearest barrier for obtaining patent protection since many inventions in this area solely involve the discovery of genomic DNA or specific biological functions. Thus, in order to facilitate the grant of patents, it is in the distinction between discoveries and inventions that traditional patent law conceptions have been more significantly modified to accommodate the needs of inventors in gene technologies. Both the requirement for an invention and the exclusion of discoveries have been revised to a point at which the notion of invention has become blurred and now invades the subject matter traditionally included within the concept of discovery.[71] In fact, the scope of the definition of invention has been expanded to cover, under certain conditions, inventions concerning naturally occurring human genes.

5.3.2 The European Approach towards the Patentability of Discoveries Concerning Human Genetic Material

The blurred line between non-patentable discoveries of human genetic substances and technical inventions from such basic raw material has been addressed by the EPO by applying the so-called isolation principle. This approach allows the patentability of naturally occurring substances once they have been isolated by man using technical means. In the case of human genetics, the isolation logic makes it possible to transform the discovery of human DNA into an invention by isolating a specific gene or gene sequence.

Even before the First Directive Proposal in 1988, the EPO had recognized isolation as the step giving technical character to discoveries of natural elements. The 1987 EPO Examination Guidelines did not make specific mention of inventions concerning DNA or other elements of the

---

[70] Ibid 253.

[71] Westerlund (n 13) 23 and 36; Denis Schertenleib, 'The Patentability and Protection of DNA-Based Inventions in the EPO and the European Union' (2003) 25 EIPR 125. Nonetheless, this new approach has not ceased to be controversial. It has raised fears that patents will be granted to fundamental discoveries, thus having an unwarranted impact on biomedical innovation at large, see Conley and Makowski (n 63). Concerns against excessive ownership over upstream inventions are discussed in Chapter 7.

human body but, with regard to naturally occurring substances in general, they provided that:[72]

> If a man finds out a new property of a known material or article, that is mere discovery and unpatentable. If however a man puts that property to practical use he has made an invention which may be patentable. ... To find a substance freely occurring in nature is also mere discovery and therefore unpatentable. However, if a substance found in nature has first to be isolated from its surroundings and a process for obtaining it is developed, that process is patentable.

Therefore, the EPO admitted the patentability of products or substances that were previously occurring in a mixture of natural elements but had not been identified, by understanding that in these cases the invention resides on the identification of the substance and isolation for useful purposes in a usable or pure form in which it did not exist in nature.[73] This approach was again emphasized in the 1988 Joint Statement of the USPTO, EPO and JPO, which noted that:[74]

> Purified natural products are not regarded under any of the three laws as products of nature or discoveries because they do not in fact exist in nature in an isolated form. Rather, they are regarded for patent purposes as biologically active substances or chemical compounds and eligible for patenting on the same basis as other chemical compounds.

However, although the EPO had established a fairly clear doctrine on the patenting of isolated natural substances, it had not addressed the issue of human body elements, and nor had the EPO included the isolation approach in the texts of the EPC or the Implementing Regulations. This question was only explicitly addressed in the EPO Examination Guidelines. The solutions provided in the Guidelines offered valuable guidance for the examining bodies at the EPO, and seemed to meet many of the needs of applicants. Nonetheless, the Guidelines' impact was limited by the fact that they are neither binding on the Board of Appeals of the EPO,

---

[72] EPO Examination Guidelines of 1987 (n 54) Part C-IV, 2.3. A similar statement is contained in the current version of the EPO Examination Guidelines (n 20) Part G-II, 3.1.

[73] See First Directive Proposal (n 22) 41 (reasoning Article 8).

[74] 1988 Joint Statement of the USPTO, EPO and JPO, quoted in R Stephen Crespi, 'Patents on Genes: Can the Issues Be Clarified?' (1999) 3 BIO-Science Law Review 199.

deciding in the final instance on patentability, nor on national courts competent in nullity procedures regarding European patents.[75] As a result, the EPO patent grant practice and the Examination Guidelines were developing on a case-by-case basis that reflected the immediate needs of the Examining Division and did not deal with all the problems in this area or did not do so in an exhaustive manner.[76] There was an almost complete lack of any reliable legislative guidance on the distinction between discoveries and inventions in the field of human genetics that could respond to the concerns of the European biotech industry.[77] To address this issue, the Biotech Directive included specific provisions interpreting the distinction between discoveries and inventions and the isolation doctrine for inventions related to the human body.

The Biotech Directive fully applied the well-known distinction between discoveries and inventions that the EPO had set with regard to biotechnology. It abides by the limitations existing under the relevant provisions of the EPC and the national patent laws of the Member States, and it is therefore primarily based, among others, on the assumption that discoveries as such are not regarded as patentable inventions.[78] With regard to human genetic material, Article 5 of the Biotech Directive was drafted to clarify the distinction between discoveries and inventions in relation to human body elements. The Biotech Directive confirmed the EPO practice of excluding naturally occurring elements as they are found in nature and prohibited at first instance the patentability of human genes in their natural status.[79] It excluded human genes in their natural form

[75] First Directive Proposal (n 22) point 33. The EPO Examination Guidelines contain detailed instructions for the personnel of the EPO on the practice and procedure to be followed in the various aspects of the examination (e.g. formal examination, search, substantive examination, opposition and other procedures, etc.) of European applications and patents in accordance with the EPC and its Implementing Regulations.

[76] First Directive Proposal (n 22) point 44.

[77] See Ibid, point 27, page 13 explaining that the existing legal framework for protecting biotechnological inventions in the Member States did not satisfy either the needs of science or industry in this field. See also Stephen Holmes, 'The New Biotech Patents Directive' [1998] Patent World 16.

[78] See First Directive Proposal (n 22) points 27 and 42.

[79] Directive 98/44/EC of the European Parliament and of the Council of 6 July 1998 on the legal protection of biotechnological inventions [1998] OJ L213/13, Recital 13 and art 5(1).

because they constitute mere discoveries, but also in response to fundamental principles protecting the dignity and integrity of the person.[80] Moreover, this exclusion also responded to fears that allowing patents on life would lead to the undue restriction of individual rights, and would entail the risk of incalculable and excessive dependence.[81]

However, after excluding from patentability human genes in their natural form, the Biotech Directive then reaffirms in Article 5(2) the long-standing practice of the EPO that naturally occurring substances are considered to be patentable inventions provided that they are isolated from their surroundings.[82] The Commission commentary is relevant in this regard to Article 2 of the 1992 Amended Proposal, which clarifies that only 'parts of the human body' per se, which means parts of the human body as found inside the human body, are unpatentable.

The Commission reemphasized this approach in its 2005 report to the European Parliament and the European Council pursuant to Article 16c of the Biotech Directive.[83] In this report the Commission supported the

---

[80] Ibid, Recital 16. See Commission proposal for a European Parliament and Council Directive on the legal protection of biotechnological inventions (1996) [COM(95) 661 final] OJ C296/4, Recital 13; Report on the proposal for a European Parliament and Council Directive on the legal protection of biotechnological inventions (1997) [COM (95)0661 – C4-0063/96 – 95/0350(COD)] (Parliament Report 1997) Amendment 11 (Recital 13). See also Samantha A Jameson, 'A Comparison of the Patentability and Patent Scope of Biotechnological Inventions in the United States and the European Union' (2007) 35 AIPLA Quarterly Journal 193.

[81] Herdegen (n 10).

[82] Biotech Directive (n 79) Recitals 20 and 21 and art 5(2). See also First Directive Proposal (n 22) Articles 8 and 9; 1996 Amended Proposal (n 80) Recitals 15 and 16 and Article 3; Parliament Report 1997 (n 80) Amendments 14 and 15 (Recitals 15 and 16), Amendment 16 (introducing new Recital 16b), and Amendment 49 (Article 3). See also Nick Scott-Ram, 'Biotechnology Patenting in Europe: The Directive on the Legal Protection of Biotechnological Inventions: Is This the Beginning or the End?' [1998] BIO-Science Law Review 43; Trevor Cook, 'The Human Genome Project: Crucial Questions the Biotechnology Directive Does Not Answer' [2000] European Lawyers; Leslie G Restaino, Steven E Halpern and Eric L Tang, 'Patenting DNA-Related Inventions in the European Union, United States and Japan: A Trilateral Approach or a Study in Contrast?' (2003) 2003 UCLA JL & Tech 2.

[83] Article 16(c) of the Directive provides that each year the Commission must transmit a report to the European Parliament and the Council on the development and implications of patent law in the field of biotechnology and genetic engineering. See EC Report from the Commission to the Council and the European Parliament: 'Development and Implications of Patent Law in the Field

logic set out in the text of Article 5(2) of the Biotech Directive and clarified that as explained in Recital 21 the reasoning is that, to qualify for patentability, an element from the human body including a sequence or partial sequence of a gene, must for instance be the result of technical processes that have identified, purified, characterized and multiplied it outside the human body and that cannot be found in nature.

A reading of Articles 5(1) and 5(2) together explains how the principles of Article 52 EPC are to be applied in the case of inventions related to human genes. The Biotech Directive allows the patenting of the human gene sequences if they are isolated from their natural surrounding, 'even if the structure of that element is identical to that of a natural element'.[84] Such interpretation confirms that the results of genomic research can be patented if they are isolated and other conditions are met, although the rights conferred by the patent do not extend to the human body and its elements in their natural environment.

The exclusion from patentability of the human body and its parts is also in line with UNESCO's view that 'the human genome in its natural state shall not give rise to financial gains',[85] which implicitly refers to the distinction between patentable inventions and simple discoveries as a fundamental distinction in the context of an increasing demand for patents in the field of human genetics.[86] Furthermore, this practice is also consistent with the 1997 Council of Europe Convention for the Protection of Human Rights and Dignity of the Human Being with regard to the Application of Biology and Medicine.[87]

Articles 5(1) and (2) of the Biotech Directive were incorporated into new Rules 23e(1) and (2) (now Rules 29(1) and (2)) of the EPC

---

of Biotechnology and Genetic Engineering' [COM(2002) 545 final] (First Commission Report 2002) 17.

[84] Biotech Directive (n 82) Recital 20 and art 5(1) and (2). See also Andrea D Brashear, 'Evolving Biotechnology Patent Laws in the United States and Europe: Are They Inhibiting Disease Research' (2001) 12 Ind Int'l & Comp L Rev 183.

[85] UNESCO Universal Declaration on the Human Genome and Human Rights of 11 November 1997, art 4.

[86] Biotech Directive (n 82) Recital 16; Noëlle Lenoir, 'Patentability of Life and Ethics' (2003) 326 Comptes Rendus Biologies 1127. See also Giovanni Macchia, 'Patentability Requirements of Biotech Inventions at the European Patent Office: Ethical Issues' in Roberto Bin et al (eds), *Biotech Innovations and Fundamental Rights* (Springer 2012).

[87] Convention for the Protection of Human Rights and Dignity of the Human Being with regard to the Application of Biology and Medicine: Convention on Human Rights and Biomedicine of 4 April 1997, arts 1, 2 and 21.

Implementing Regulations in 1999,[88] which reproduce the same wording laid out in the Biotech Directive. The EPO Guidelines for Examination were also amended to introduce the Biotech Directive's provisions concerning the patentability of human body parts in the light of the exclusion of discoveries. In this regard, new section 2a was incorporated in Part C-IV (which deals with issues of patentability) of the 2003 Revised Guidelines.[89]

5.3.3 The EPO's Implementation of the Isolation Doctrine

As can be seen from the previous section, the isolation principle has been part of the EPO practice since before the Biotech Directive was adopted in 1998. Some examples of patents granted over inventions relating to isolated human DNA are:

- EP0041313 granted in 1990 regarding DNA sequences, recombinant DNA molecules and processes for producing human fibroblast interferon
- EP0101309, granted in September 1991 relating to a DNA fragment encoding human H2-preprorelaxin or
- EP0630405 relating to DNA sequences and novel seven transmenbrane receptors, which was granted in April 1998 but revoked in June 2001.

In particular, the question of whether isolated human genetic material could be patentable was debated before the OD of the EPO in the Relaxin case.[90] The patent was challenged among other grounds under Article 52(2)(a) EPC for not being concerned with an invention but with a discovery. The appellants claimed that the essence of the invention, which was elucidation of the genetic sequence of the H2-relaxin gene, was no more than a discovery of the characteristics of a substance that had probably existed in nature for thousands of years. So, in the meaning of the provision of Article 52(2)(a) EPC, the patent related to a discovery and was consequently not patentable.

---

[88] EPO Decision of the Administrative Council of 16 June 1999 amending the Implementing Regulations to the European Patent Convention (OJ EPO 1999, 437).

[89] Guidelines for Examination of the EPO revised in 2003, Part C-IV, 2a. This section is now Part G-II, 5.2 of the EPO Examination Guidelines (n 20).

[90] *Relaxin/HOWARD FLOREY INSTITUTE* (T 0272/95) [2002] (EPO (OD)).

However, the patentees argued that according to new Rule 23(e)(2) EPC introducing the Biotech Directive, it was clear that patent protection should extend to elements isolated from the human body or otherwise produced by means of a technical process even if the structure of that element was identical to that of a natural element. In line with the patentee's argument, the OD reemphasized the long-standing practice of the EPO and stated that the claimed DNA fragments encoding the human protein preprorelaxin or to the human preprorelaxin per se, had been obtained by technical processes,[91] were not to be considered as discoveries, and therefore, did not fall within the category of unpatentable inventions.

More recent cases such as ICOS Corporation, University of Utah, Zymogenetics or Eli Lilly have also acknowledged the patentability of isolated human genes. For example, in the case of the University of Utah's breast and ovarian cancer patent application, the main claim related to 'an isolated nucleic acid which comprises a coding sequence for the human BRCA1 polypeptide, wherein said polypeptide … comprises the amino acid sequence of SEQ ID NO: 82 …'.[92]

The patent was granted in November 2001 and opposition proceedings were immediately filed with some opponents arguing that the patent was against Article 52(2)(a) of the EPC since the claimed genes were discoveries and not patentable. However, the OD rejected this objection and applied the logic laid out in Rule 29(2) (former Rule 23(e)(2)) of the EPC. On appeal, the TBA at the EPO also confirmed the previous EPO practice in the Relaxin case as well as the compliance of the isolated nucleic acid probes claimed by the applicant with Rule 29(2) of the EPC.[93]

Another example is the case of Eli Lilly where the appellant (HGS) withdrew all its previous requests and filed a new main request with the word 'isolated' in its main claim, 'an isolated nucleic acid molecule comprising a polynucleotide sequence encoding a Neutrokine-α polypeptide …'.[94]

---

[91] Ibid, points 7 and 8.

[92] *Breast and ovarian cancer/ UNIVERSITY OF UTAH* (T 1213/05) [2007] (EPO (TBA)) point VII.

[93] *University of Utah* (n 92) points 43, 44, and 45.

[94] European Patent EP-B-0939804, Neutrokine alpha (1996). See also *Neutrokine alpha/HUMAN GENOME SCIENCES* (T 0018/09) [2009] (EPO (TBA)) point XIX.

5.3.4 Objections To and Justifications For Patenting Isolated Human Genes

5.3.4.1 Arguments against the isolation approach

Despite existing arguments supporting the isolation principle, the application of this approach in the patenting of human genetic material has also raised concerns in different social groups, including claims that the isolation doctrine overrides the patent law distinction between discoveries and inventions.[95] In this regard, critics and objections to the isolation principle can be reduced to the question of whether isolation is enough to transform natural elements into man-made inventions.

5.3.4.1.1 Natural and isolated genes are identical The most frequent claim against the grant of patents over isolated genetic material is that the resulting isolated substance is identical to, and has the same properties as such element in its naturally occurring status. The argument is that isolated, purified and synthesized human genes maintain identical or very similar characteristics to those in their natural form and realize the exact same function.[96] The resulting products do not change substantially from what exists in nature, except for purity and stability.[97] That is, genetic substances in their natural and purified versions are essentially indistinguishable and do the same thing. Moreover, the fact that the process for isolating a gene sequence or protein involves man-made techniques that cannot be found in nature[98] is not relevant since the subject of the patent is the isolated human genetic material and not the technique that is employed to obtain it.

5.3.4.1.2 Trivial human intervention Another very frequent objection is based on the argument that the intervention of human intelligence in

---

[95] Amanda Odell-West, '"Gene"-Uinely Patentable? The Distinction in Biotechnology between Discovery and Invention in US and EU Patent Law' [2011] IPQ 304; Tine Sommer, *Can Law Make Life (too) Simple?: From Gene Patents to the Patenting of Environmentally Sound Technologies* (1st edn, DJØF Publishing 2013) 251.

[96] Conley and Makowski (n 63); Nuno Pires de Carvalho, 'The Problem of Gene Patents' [2004] 3 Wash U Global Stud L Rev 701; Holman (n 12); Pires de Carvalho, *The TRIPS Regime of Patent Rights* (2010) (n 55) 261; Sommer (n 95) 255.

[97] Westerlund (n 13) 43; Cornish, Llewelyn and Aplin (n 33) 926.

[98] See Palombi, 'Patentable Subject Matter, TRIPS and the European Biotechnology Directive: Australia and Patenting Human Genes' (n 58); Sommer (n 95) 519.

the isolation process is trivial. In this regard, it has been claimed that man does not create genes or their corresponding proteins by isolating them, but simply discovers them.[99] Genes are nature-made substances that are unique in the sense that scientists cannot artificially create them to produce identical results such as coding for the same protein.[100] Thus, individuals may alter, modify, isolate or interfere with the natural properties of genetic material in different ways, but cannot possibly invent and produce DNA. Furthermore, the value of the techniques involved in the isolation process has also been questioned. It has been argued that the process of isolating human genetic material does not entail too much intellectual effort or ingenuity but that isolation is now a routine process carried out on a daily basis.[101] Therefore, a significant level of human intervention cannot be found in the isolated DNA substances or in the process for isolating them.

According to this view, granting patents over isolated genes or proteins seem to be too much of a reward for bioengineers. It has been argued that the reward of being granted a patent should be proportional to the expended efforts rather than to the contribution that has been made – for example, in spite of the fact that isolated gene sequences can be extremely important for the development of cures for diseases affecting people worldwide – or the money that has been invested; thus awarding a patent for easily isolated natural substances would be excessive.[102] Furthermore, as biotechnological developments are mostly based on previous knowledge and research outcomes, it may seem unfair that the final contributor in the research chain, who 'only' isolates a certain substance, gets such a reward.

---

[99] Ned Hettinger, 'Patenting Life: Biotechnology, Intellectual Property, and Environmental Ethics' (1994) 22 BC Envtl Aff L Rev 267; Sena (n 13); Amanda Odell-West, '"Gene"-Uinely Patentable? The Distinction in Biotechnology between Discovery and Invention in US and EU Patent Law' (n 95).

[100] Patricia A Lacy, 'Gene Patenting: Universal Heritage vs. Reward for Human Effort' (1998) 77 Or L Rev 783; Pires de Carvalho, *The TRIPS Regime of Patent Rights* (2010) (n 55) 253 and 260.

[101] Mark A Chavez, 'Gene Patenting: Do the Ends Justify the Means' (2002) 7 Computer L Rev & Tech J 255. The use of DNA for analysis or manipulation usually requires that it is isolated and purified to a certain extent, see Keith Wilson and John M Walker (eds), *Principles and Techniques of Biochemistry and Molecular Biology* (7th edn, Cambridge University Press 2009) 164.

[102] Hettinger (n 99); Lori B Andrews and Jordan Paradise, 'Gene Patents: The Need for Bioethics Scrutiny and Legal Change' (2005) 5 Yale J Health Pol'y L & Ethics 403.

5.3.4.1.3 Discrimination in favour of human DNA patents Finally, given the simplicity of the isolation process, equating the extraction of a gene from its natural environment with inventions having technical character raises questions about the compatibility of such approach with Article 27(1) of the TRIPS Agreement, which requires patent protection for all types of inventions under the same conditions.[103] Nonetheless, the TRIPS Agreement is aimed at establishing minimum standards that countries may modify to accommodate the needs of a particular industry.

5.3.4.2 Arguments supporting the isolation approach

Notwithstanding existing concerns, the reality is that biological sciences, especially gene technologies, have evolved to a point where it is increasingly difficult to set the barrier between upstream research, that is more related to discoveries, and downstream inventions. As a result, the patentability of isolated genetic material has gained growing acceptance among private and public stakeholders.

5.3.4.2.1 Substantial change between genes in their natural status and their isolated form The first argument counteracting critics to the isolation doctrine is that there is a substantial difference between DNA sequences as they exist in the body and their isolated version. The process of isolating allows a different degree of accessibility to the desired substance[104] in the sense that an element from the human body is physically removed, making it available for the first time in a tangible incarnation outside the human body.[105] The resulting invention derives from natural substance but it is man-made and could not possibly come into existence without human intervention. Therefore, all products based on isolated genes would be the result of human activity, since such genes did not exist in their isolated form prior to their creation in the laboratory.

On the other hand, it can be noted that the isolated gene and its natural precursor are not fully identical. Genes in the body have both coding and non-coding regions, while isolated genes have been manipulated to eliminate the non-coding region but apparently still continue to perform

---

[103] TRIPS Agreement (n 2) art 27(1). See also Palombi, 'The Impact of TRIPS on the Validity of the European Biotechnology Directive' (n 5); Kevin Emerson Collins, 'Propertizing Thought' (2007) 60 SMU L Rev 317.

[104] Beier, Crespi and Straus (n 17) 55.

[105] Amanda Warren-Jones, *Patenting rDNA: Human and Animal Biotechnology in the United Kingdom and Europe* (Lawtext Pub 2001) 175–176. See also Konrad B Becker, 'Are Natural Gene Sequences Patentable?' (2000) 73 Int Arch Occup Environ Health (Supp) S19.

the same function as a naturally occurring gene.[106] By eliminating the parts that do not carry any relevant genetic information, something new is added through the isolation process.[107] The composition of the resulting product is thus different and encompasses a technical teaching that is acquired due to human intervention.

The argument that natural DNA and isolated parts of it are not the same has found the strongest support within the context of cDNA.[108] Patent applications do not usually claim genomic DNA but inventions related to cDNA, which is a totally artificial construct.[109] cDNA is the result of isolating and copying the mRNA of an organism into a mirror image of itself.[110] The resulting cDNA is thus an artificial complementary copy of the original DNA, which possesses the same sequence of nucleic acids as the original piece of DNA in the organism's genome. All cDNA is derived from and through laboratory procedures applied to compounds that do exist in nature. However, although a cDNA sequence's structure is a copy of a naturally occurring mRNA molecule,[111] the reality is that obtaining mRNA from DNA and translating it into cDNA involves a technical laboratory process that creates a new substance that is artificial and without DNA non-coding regions, and thus not the same as that which occurs in nature.[112] In this regard, since most inventions over isolated genes cover ESTs, which are parts of artificially created cDNA, according to this view they cover something strictly different than the naturally occurring genomic DNA.

---

[106] Andrews (n 31); Conley and Makowski (n 63); Andrews and Paradise (n 102).

[107] Chavez (n 101); Schertenleib, 'The Patentability and Protection of DNA-Based Inventions in the EPO and the European Union' (n 71).

[108] See the US Supreme Court decision in *Association for Molecular Pathology v Myriad Genetics, Inc.*, 569 US (2013) and the subsequent decision of the High Court of Australia in *D'Arcy v Myriad Genetics Inc.* [2015] HCA 35.

[109] R Stephen Crespi, 'Patents on Genes: Clarifying the Issues' (2000) 18 Nat Biotech 683.

[110] Eric S Lander and others, 'Initial Sequencing and Analysis of the Human Genome' (2001) 409 Nature 860; J Craig Venter and others, 'The Sequence of the Human Genome' (2001) 291 Science 1304.

[111] James Bradshaw, 'Gene Patent Policy: Does Issuing Gene Patents Accord with the Purpose of the U.S. Patent System' (2001) 37 Willamette Law Review 637.

[112] Conley and Makowski (n 63); Oliver Mills, *Biotechnological Inventions, Moral Restraints & Patent Law* (Revised Edition, Ashgate 2010) 143.

5.3.4.2.2 Difficulties in the isolating process On the other hand, isolating a gene sequence is not a straightforward process but involves great effort as well as man-made technical processes and tools. The difficulty lies in identifying that a particular gene that may have a valuable function like reducing the speed of tumour growth and spread, using information from computer analysis and laboratory tests.[113] Moreover, it must be noted that measures for obtaining sequences, no matter how simple they might be, are of a technical nature.[114] Furthermore, the skilled person can also encounter important difficulties in isolating the product, such as the trials and errors required and the thorough research pursued in order to isolate and analyse the structure and function of the product in question.[115] Therefore under this view, granting patents to those biotech engineers that have spent their time and effort in identifying, isolating or purifying DNA sequences so as to make them available in a different form would be a proportional reward.

5.4 THE 'ISOLATION PLUS INDUSTRIAL APPLICABILITY' DOCTRINE

Given the concerns regarding the number of patents over newly discovered human genetic material without a known practical applicability, the drafters of the Biotech Directive felt it necessary to revise the distinction between discoveries and inventions in the realm of human gene patents. In this regard, next to the requirement of isolation, the Biotech Directive introduced the explicit requirement of disclosing the industrial application of inventions concerning isolated human gene sequences in the patent application. The implications of this new requirement on the interpretation of the industrial applicability of human gene sequences have been examined in Chapter 4. In contrast, this chapter focuses on the implications of Article 5(3) of the Biotech Directive on the interpretation of the exclusion from patentability of human genetic discoveries.

The need of reducing a discovery to practice and make it useful in order to become an invention has continuously been part of the EPO practice;[116] however it did not have any explicit reflection in existing

[113] Bradshaw (n 111).

[114] Andreas Oser, 'Patenting (partial) Gene Sequences Taking Particular Account of the EST Issue' (1999) 30 IIC 1.

[115] Sven Bostyn, 'The Patentability of Genetic Information Carriers' [1999] IPQ 1.

[116] See EPO Examination Guidelines of 1987 (n 54) Part C-IV, 2.3.

legislation. In this regard, what the Biotech Directive does in Article 5(3) is basically to further refine the requirement of Article 3, which recognizes that isolated biological material is not a discovery and can be patented, in the context of naturally occurring human genetic material by imposing the additional requirement of disclosing the industrial applicability for isolated sequences or partial sequences of human genes for these substances to acquire the character of invention.

While the Biotech Directive essentially confirms the EPO practice according to which isolated human material is patentable, it demarcates the line between discoveries and inventions more clearly in this context.[117] Recital 34 of the Biotech Directive contains a conservative note stating that its text does not intend to change the meaning of discovery and invention developed by national, European and international patent law. However, with regard to human gene inventions, Article 5(3) adds important features to such definitions. Nonetheless, as indicated by Recitals 22 to 24, the requirement of specifying a technical function would only be applicable when gene sequences or partial sequences are claimed 'as such'.

In respect of the importance of the notion of usefulness in the distinction between inventions and discoveries concerning the human body, the First Directive Proposal explained that, although an invention may involve a naturally occurring substance, there would be a considerable difference between the product as it existed in nature and the product in a useful form.[118] Subsequently, in order to address European Parliament's concerns[119] regarding the need to clarify the difference between discoveries and inventions where the patentability of elements of human origin is concerned, the 1997 Amended Proposal[120] as well as the final text kept the provisions on the additional requirement for isolated human gene sequences of disclosing their industrial application. In particular, even when the 1997 Parliament's report did not specifically discuss the role of industrial application in acquiring technical character,

---

[117] See Sven J R Bostyn, 'A Decade After the Birth of the Biotech Directive: Was It Worth the Trouble?' in Emanuela Arezzo and Gustavo Ghidini (eds), *Biotechnology and Software Patent Law* (Edward Elgar 2011) 223.

[118] First Directive Proposal (n 22) 42 (reasoning Article 8).

[119] Parliament Report 1997 (n 80) Amendment 16b (Recital 16b): 'Whereas a mere sequence of DNA segments does not contain a technical teaching and is therefore not a patentable invention'.

[120] Amended Proposal for a European Parliament and Council Directive on the legal protection of biotechnological inventions (1997) [COM(97) 446 final] 1.

Recital 16b of the 1997 Amended Proposal (incorporating amendment 16b of the 1997 Parliament report) made it clear that having a practical utility is what gives technical character to an isolated gene sequence.

The Biotech Directive's intention was to impose an additional requirement for isolated gene sequences so that isolating them is no longer enough to acquire technical character, but the elucidation of a practical utility is also necessary to become an invention in patent law terms. Thus, while a mere discovery is not patentable,[121] an isolated gene sequence with a function contains a technical teaching and can therefore be patentable.[122] As a consequence of such an understanding, the requirements of technicality, practicability and industrial applicability as preconditions for patentability, which are in principle different and independent, seem to essentially overlap with each other when assessing the eligibility of human genetic discoveries for patent protection.

5.4.1 Industrial Applicability as a Pre-Condition for Acquiring Technical Character

The obligation of Article 5(3) has been generally understood as an additional requirement to the isolation principle for certain types of human body inventions. In these cases, being industrially applicable is what marks the distinction between the teaching of a mere reproduction of genetic information (discovery) and the indication of a function of a DNA sequence contributing to a technically exploitable result (invention).[123] If the claimed product is restricted to the mere reproduction of genetic information, it is merely an enrichment of the state of knowledge, a pure discovery; however, if as a result of the indication of the function, a claimed DNA sequence contributes to a technically exploitable result, it is an invention.[124] Therefore, a simple isolated gene cannot be patentable until it is usable in some specific industrial application.[125] Isolation and function are the two key operative concepts for the patentability of human genes under the Biotech Directive. In this regard, a principal question is to determine the relationship between technical character, isolation and industrial applicability in these types of inventions.

---

[121] Biotech Directive (n 82) Recital 16.
[122] Ibid, Recital 23.
[123] Oser (n 114).
[124] Ibid.
[125] Sena (n 13).

Articles 5(2) and (3) of the Biotech Directive equate isolated naturally occurring elements with an indication of function with elements produced as a result of a technical process. According to the Biotech Directive, the mere discovery of the base pair sequence of a gene is not an invention; however, locating and isolating a previously unknown gene, determining its function and making it accessible for further exploitation is a technical solution to a technical problem.[126] By rejecting applications over gene sequences with unknown uses on the basis that they are not really inventions,[127] the Biotech Directive also addresses the fact that all genes that are discovered have to be isolated first by various technical methods.[128] Thus, the distinction between discoveries and inventions in this field is to be found somewhere else, namely, in the identification of a practical utility.

It seems clear that the Biotech Directive relies on the notions of utility and function and their implications in the distinction between discoveries and inventions. Nevertheless, there has been much debate regarding what the correct interpretation for Article 5(3) should be. Recital 23 of the Biotech Directive states that a simple DNA section without indication of a function does not contain a technical teaching and hence does not represent a patentable invention. Such wording contradicts the EPO previous practice by equating technical teaching with patentable, while having a technical teaching is a pre-condition for being an invention, and not necessarily a patentable one. This would lead to the conclusion that an isolated gene without an indication of function would be an invention although not patentable even though it does not have technical character.

Due to this (apparent) contradictory provision, for some, isolation is enough for a discovered DNA sequence to provide a technical teaching, and thus viewing Article 5(3) as an additional ingredient for an isolated gene to become an invention would erroneously imply that Article 57 EPC would be an objection to the patentability of mere discoveries,

---

[126] Schertenleib, 'The Patentability and Protection of DNA-Based Inventions in the EPO and the European Union' (n 71); Herbert Zech and Jürgen Ensthaler, 'Stoffschutz bei gentechnischen Patenten – Rechtslage nach Erlass des Biopatentgesetzes und Auswirkung auf Chemiepatente' [2006] GRUR 529.

[127] Ng-Loy Wee Loon, 'Patenting of Genes – A Closer Look at the Concepts of Utility and Industrial Applicability' [2003] IIC 393.

[128] Schertenleib, 'The Patentability and Protection of DNA-Based Inventions in the EPO and the European Union' (n 71); Amanda Odell-West, '"Gene"-Uinely Patentable? The Distinction in Biotechnology between Discovery and Invention in US and EU Patent Law' (n 95).

rather than a patentability requirement for inventions.[129] Moreover, doing so could also lead to the conclusion that if a patent application over a gene sequence is found to be an invention, it should automatically be deemed capable of industrial application.

The difficulty in achieving a clear conclusion on the status of Article 5(3) is reflected in the ICOS Corporation case where the OD at the EPO contradicted itself by first saying with regard to patentability that the claimed isolated gene was not a discovery and then that there was no technical character.[130] However, in spite of the at times contradictory interpretation of Article 5(3), the prevailing view is that the provisions of Article 5 must be interpreted altogether as an integral requirement for human gene patents. In this regard, interpreting Article 5(3) as not essential for the distinction of discoveries and inventions over human genes would ultimately misunderstand the Biotech Directive's aim of imposing the obligation to elucidate the function of an isolated gene as a precondition for obtaining technical character and thus be an invention.

5.4.2 The Implementation of the Isolation Plus Industrial Applicability Approach

In order to discover a gene, this must be first isolated. Therefore, considering merely isolated human gene sequences without industrial applicability as possible patentable inventions seems arguable since it is how the discovery is applied that turns it from a pure discovery into a real benefit to mankind worthy of protection as an invention.[131] The criterion of function adds to the invention concept the aspect of practical use, an aspect which, for newly discovered human gene fragments, may be the balanced course required for keeping the concept within the limits necessary to serve the fundamental objectives of the patent system.[132] What must be patented in every case is the set of technical instructions

---

[129] See Amanda Odell-West, '"Gene"-Uinely Patentable? The Distinction in Biotechnology between Discovery and Invention in US and EU Patent Law' (n 95).

[130] *Novel V28 Seven Transmembrane Receptor/ICOS CORPORATION* [2002] OJEPO 293 (EPO (OD))11(i) and (ii).

[131] See Holmes (n 77).

[132] See Westerlund (n 13) 44. See also Pila, 'The Future of the Requirement for an Invention: Inherent Patentability as a Pre- and Post-Patent Determinant' (n 36) 57–58.

that inventively solve a particular technical problem.[133] Thus, an examination of paragraphs 1 and 2 of Article 5 of the Biotech Directive does not itself permit a clear delineation between discovery and invention since a merely isolated gene does not provide any technical, practically applicable solution for a technical problem.

The value of a patent mainly depends on the applicability of the associated invention. In this regard, a vertical test for technical character would only allow 'knowledge ready to be carried out by any relevant person skilled in the art' to be patented, while preventing other subject matter from being patented, such as mere descriptions of observations.[134] Technical knowledge is something that a person skilled in the art can perform, as opposed to theoretical knowledge,[135] such as isolated human DNA fragments of unknown utility. Therefore, the contribution to the art on which gene patents are based is the gene's availability in a form that can be utilized by many, and not only by the inventor.

In fact the Boards of Appeal at the EPO have treated inherent patentability and industrial character as related in several cases.[136] In particular, after the Biotech Directive was adopted, the EPO decision in Multimeric highlighted the overlapping relationship between having technical character and being industrially applicable. In this case, the TBA equated being industrially applicable with not being a mere discovery of interesting research results.[137] Moreover, the Board also noted that the question of whether a claimed subject matter is industrially applicable is directly linked to the question of whether the claimed object is a mere discovery, for example, the result of purely intellectual activity with no practical or technical character.[138] Unfortunately that reasoning was not furthered since the invention was deemed susceptible of industrial application, but the decision made clear that isolation of human genetic material is not enough and that identifying a concrete industrial application is also a pre-condition for the patenting of human genes.

---

[133] Parliament Report 1997 (n 80) 33.

[134] This differs from horizontal technical character, which would differ between fields or disciplines at a peer level, see Bakels (n 27) 181.

[135] Bakels (n 27) 162–163. See also Pottage and Sherman (n 18) 153.

[136] See *Traffic Regulations/CHRISTIAN FRANCERIES* (T 0016/83) [1985] (EPO (TBA)); *Computer-Related Invention/VICOM* (T 0208/84) [1986] (EPO (TBA)); See also Pila, 'The Future of the Requirement for an Invention: Inherent Patentability as a Pre- and Post-Patent Determinant' (n 36).

[137] *Multimeric Receptors* (n 7) point 2.

[138] Ibid, point 2.

This view was also confirmed by the prosecution at the UK Court of Appeal in *Eli Lilly v HGS*, which asserted that an objection to the patentability of a gene sequence based on Article 57 of the EPC (industrial application) is in effect no different from an objection that the disputed invention is a mere discovery.[139] The same point was echoed by the Advocate General in Monsanto:[140]

> The great importance attached by Directive 98/44 to the function performed by a DNA sequence is naturally intended to permit a distinction to be drawn between 'discovery' and 'invention'. The isolation of a DNA sequence without any indication of a function constitutes a mere discovery and as such is not patentable. Conversely, the sequence is transformed into an invention, which can then enjoy patent protection, through the indication of a function that it performs.

These decisions further show how, with the adoption of the Biotech Directive, the European patent law requirement of industrial application has become an essential pre-condition for human gene discoveries to qualify as inventions. This requirement adds to the long-standing isolation principle, thus imposing an extra hurdle that patent applicants need to overcome in order to obtain patent protection for discoveries concerning sequences and partial sequences of human genes.

From an international perspective, the decisions of the US Supreme Court and the High Court of Australia in the Myriad case[141] have both emphasized the role of the utility criterion in the invention-discovery distinction. These decisions considered merely isolated genetic material containing the same information as the naturally occurring substance inside the cell, as opposed to cDNA, as informational matter rather than inventions susceptible of patent protection. While the invention claimed by Myriad was a nucleic acid isolated by human action, the courts considered that the existence of the information stored in the relevant sequences, which was substantially unaltered by the isolation process, was the essential element of the invention as claimed. Further, the courts did not find that the utility of the isolated material was sufficiently linked to the fact that the nucleic acids had been artificially isolated, and therefore, Myriad's claims to isolated DNA without any limitation to a useful application were claims to products of nature. As regards the conception of isolated genetic material as technical inventions, the

---

[139] *Eli Lilly & Co v Human Genome Sciences Inc* (EWCA) (n 94) point 60.

[140] C-428/08 *Monsanto Technology LLC v Cefetra BV and Others* [2010] ECR I-06765, Opinion of Advocate General Mengozzi, point 31.

[141] See (n 108).

arguments put forward in the Myriad decisions in Australia and the US have again shed light on the long-debated validity of patent grants on isolated discovered genetic material as such and the claims for imposing tailored requirements for these types of products to be patent eligible. In contrast with the Biotech Directive, these decisions reject the patentability of isolated genetic material when the structure of the claimed substance is identical to that of naturally occurring genetic material. However, the decisions in Myriad appear to follow the same underlying approach towards the requirement for isolated human genetic sequences as such to show a defined utility that Article 5 of the Biotech Directive sets out with regard to the patentability of human body parts. That is, while merely isolated human genetic material remains a discovered natural product, isolated sequences with an identified utility or, as the High Court of Australia held, a defined utility linked to the artificiality of the isolated material, are substantially different to the naturally occurring substance and amount to technical inventions under patent law. It seems that, notwithstanding existing differences between the Biotech Directive and the US and Australian courts' conceptions of genetic inventions, their approaches appear to also rely on the notion of utility as a cornerstone requirement for isolated human DNA to become a counterpart to an invention under patent law.

5.5 CONCLUDING REMARKS

Because biological inventions concern living matter that previously occurred in nature, the exclusion of discoveries from patentability plays a key role in the biotechnology field. In particular, the distinction between discoveries and inventions is especially difficult in the case of inventions over discovered human DNA sequences due to their intrinsic characteristics and the fact that on many occasions these substances do not produce a concrete technical effect (function) and thus do not possess technical character.

The EPC Implementing Regulations and Examination Guidelines interpret the exclusion of discoveries of natural substances within very narrow limits, only excluding matter that is actually 'freely occurring in nature'. Within the context of DNA patents, this means that as long as a gene sequence has been isolated from its natural surrounding, it can be patentable. However, Article 5 of the Biotech Directive introduced the requirement that in order to become an invention with technical character, a newly discovered gene sequence has to be isolated and its function also needs to be identified. According to Article 5 the two requirements need to be satisfied to transform a mere discovery of DNA material into a patentable invention.

This new approach, which is consistent with the EPO's traditional understanding of the requirement of technical character of an invention, has been incorporated into the EPC and now forms part of the EPO practice. In this regard, several EPO decisions have made clear that if no industrial application is disclosed, an isolated human gene sequence amounts to a mere discovery without technical character.

6. The requirement of industrial application and the determination of the scope of protection of gene patents

6.1 INTRODUCTION

With the rapid development of gene technologies, patent authorities started to encounter difficulties in applying general rules on patent scope to inventions over genetic material. Before determining the scope of a patent it is necessary to define the invention's contribution to the art. However, in the area of genetics, the proliferation of broad claims based on little prior art, such as patent applications claiming many functions for newly discovered ESTs, makes it difficult to carry out a clear delimitation of the technical contribution of inventions in this field.

Article 9 of the Biotech Directive deals with the determination of patent scope for inventions consisting of or containing genetic information by limiting the scope of protection of these types of patents to their ability to perform their function. In practice, this provision has been interpreted as establishing purpose-bound protection by reducing the patent scope awarded to the industrial application of the invention claimed by the patent owner when filing the patent application. It remains arguable whether this interpretation should be applied to all or only to some types of genetic inventions, yet such an approach would give the requirement of disclosing the industrial applicability of human DNA claims a key role in the determination of the scope of protection awarded, at the least, to those inventions concerning human genetic material.

This chapter first analyses the important role of the question of patent scope in gene patenting by exploring the economics of patent scope and the concerns around the grant of broad patents for genetic inventions. In this regard, the chapter also discusses the compatibility between the principal economic theories on patent scope and the Biotech Directive's approach towards the scope of protection of genetic inventions. The chapter then examines the interpretation of patent scope under the EPC by studying the EPC provisions on patent scope and their application to genetic inventions, and analyses the role and interpretation of disclosure requirements for determining the scope of gene patents under the EPC. Following this, the chapter focuses on Article 9 of the Biotech Directive, which deals with the scope of protection of genetic inventions, as an additional provision to the EPC articles on patent scope that are applicable to all types of inventions. It reviews the historical background of this provision, analyses its interpretation as establishing purpose-bound protection for genetic inventions, and discusses the approach taken by the Commission and the Parliament towards the implementation of this requirement. Finally, the chapter examines the CJEU decision in *Monsanto v Cefetra* interpreting Article 9 of the Biotech Directive and its impact on the interpretation of this provision and the future of gene patenting.

6.2 PATENT SCOPE OF GENETIC INVENTIONS

6.2.1 The Economics of Patent Scope

The economic significance of a patent depends on its scope. Together with length, the scope of a patent determines the limits of the monopoly power that a patent confers. In practice, patent scope relates to the flow rate of profit that the patentee can appropriate while the patent is enforceable.[1] It refers to the extent to which the patent holder may exclusively exploit or improve the subject matter covered by the patent;[2] or in other words, the scope of a patent equates to the range of products or processes for which the patent owner has the right to prevent others from making, using, selling or importing the patented invention.[3] In this regard, patent disputes often revolve around the question of how different from the protected invention a product must be in order to avoid infringing the exclusive rights of the patent owner.[4] Therefore, defining the scope of a patent is decisive for determining the extent of the rights

[1] Richard Gilbert and Carl Shapiro, 'Optimal Patent Length and Breadth' (1990) 21 The RAND Journal of Economics 106.

[2] Yusing Ko, 'An Economic Analysis of Biotechnology Patent Protection' (1992) 102 The Yale Law Journal 777.

[3] Agreement on Trade-Related Aspects of Intellectual Property Rights of 15 April 1994, art 28.

[4] Suzanne Scotchmer, *Innovation and Incentives* (MIT Press 2004) 103.

of the patentee but also the limitations that such rights impose on competitors.

The Coasian theory of property allocation suggests that, with low or nonexistent transaction costs, parties will always bargain to a Pareto-superior solution (regardless of how entitlements are initially assigned) and the size of the rights should only affect the personal gains of individual parties.[5] However, within the context of patent rights, further analysis of the Coase theorem has shown that the size of patentees' rights can make a difference in the equilibrium level of output of the bargaining parties.[6] Broad and narrow patent rights have different impacts on technological and scientific progress. Thus, in order to decide what the optimal patent scope would be in a given technological field it is important to consider the effects produced by different models of patent scope on distinct innovation landscapes.

The broader the scope of the patent, the larger the monopoly power of the patentee is, and the stronger his negotiating position. Broad patents confer power to exclude a greater number of competing goods from the market that may otherwise infringe on the patentee's rights.[7] In this regard, public policy analysis of optimal patent scope is usually focused on the trade-off between the reward for innovators and society's access to patented inventions. This trade-off between patentees and society has traditionally been solved by discussing the appropriate length of patents.[8] However, several authors have more recently acknowledged the primary role of patent scope as a key policy element in patent regimes.

On the one hand, if patent claims were unduly limited in scope they would not promote progress in the art by failing to provide adequate economic incentives to inventors.[9] But on the other hand, overly broad

---

[5] See Ronald H Coase 'The Problem of Social Cost' (1960) 3 Journal of Law & Economics 1; Harold Demsetz, 'Toward a Theory of Property Rights' (1967) 57 American Economic Review 347.

[6] Robert P Merges and Richard R Nelson, 'On The Complex Economics of Patent Scope' (1990) 90 Columbia Law Review 839.

[7] Ibid.

[8] See for example William Nordhaus, *Invention, Growth and Welfare: A Theoretical Treatment of Technological Change* (Cambridge, MA: MIT Press 1969); William Nordhaus, 'The Optimum Life of a Patent: Reply' (1972) 62 American Economic Review 428; Frederic M Scherer, 'Nordhaus' Theory of Optimal Patent Life: A Geometric Reinterpretation' (1972) 62 American Economic Review 422. See also Gilbert and Shapiro (n 1).

[9] Kenneth Sibley, 'Disclosure Requirements' in Kenneth Sibley (ed), *The Law and Strategy of Biotechnology Patents* (Butterworth-Heinemann 1994) 118.

patents would improperly dominate a certain area of research[10] and probably reduce the social benefits of patented inventions since increasing the breadth of a patent is typically increasingly costly in terms of the deadweight loss as the patentee's market power grows.[11] Thus, patent scope discussions usually concentrate on two main concerns. First, there is the idea that because patentees must disclose their inventions they should receive patent protection sufficient to prevent competitors from easily inventing around the inventions, and so that their patents are economically valuable.[12] However, if a technological area presents many overly broad patents, this may discourage competitors from investing in making improvements.[13] Thus a second concern is that patents should be limited in scope to motivate other inventors to continue investing in research without fear of infringing on a previous patent.[14] This follows the need to achieve a balanced definition of patent scope that rewards innovators in exchange for a minimum welfare loss. Nonetheless, reconciling these two objectives is often difficult, especially in pioneer and fast-developing fields such as genetics.

6.2.2 Broad Patents and the Case of Pioneer Innovations

Patent claims that cover things significantly different from what the patentee actually invented can result in overly broad patents covering an area of innovation that is too large rather than a specific invention. This type of claiming may be problematic since it often seeks to control either an overly large part of the existing technological landscape, that is, wide

---

[10] Ibid 118.

[11] Merges and Nelson, 'On The Complex Economics of Patent Scope' (n 6); Sven JR Bostyn, *Enabling Biotechnological Inventions in Europe and the United States: A Study of the Patentability of Proteins and DNA Sequences with Special Emphasis on the Disclosure Requirement* (European Patent Office 2001) 42–43.

[12] Alison E Cantor, 'Using the Written Description and Enablement Requirements to Limit Biotechnology Patents' (2000) 14 Harvard Journal of Law & Technology 267. See also Sven JR Bostyn, 'A European Perspective on the Ideal Scope of Protection and the Disclosure Requirement for Biotechnological Inventions in a Harmonized Patent System: The Quest for the Holy Grail' (2002) 5 J World Intell Prop 1013.

[13] Bostyn, 'A European Perspective on the Ideal Scope of Protection and the Disclosure Requirement for Biotechnological Inventions in a Harmonized Patent System: The Quest for the Holy Grail' (n 12).

[14] Cantor (n 12).

patent breadth which refers to the range of currently foreseeable commercial products that the patentee intends to claim;[15] or too many subsequent technologies, and thus, large patent depth, which refers to the currently unforeseeable range of products that the patentee intends to cover;[16] or both.

Discussions around the impact of broad scope patents are frequent within the context of pioneer innovations, like groundbreaking inventions belonging to fields the boundaries of which have not yet been determined such as genetics and other advanced biotechnologies. In this sense, there are different views regarding the question of how large the scope of patent protection for pioneer inventions should be. Nonetheless the predominant argument is that meritorious pioneer inventions that open up a new scientific field or start new industries (e.g. first genetic inventions within the context of recombinant DNA technology) are worthier of broader patent protection than more modest inventions or improvements on existing ideas.[17] This approach relies on the policy-based rationale that if pioneer inventors in new fields do not receive enough protection, they will be unable to capture adequate returns from their inventions,[18] while new entrants into the field will be discouraged.[19] Moreover, to some extent, broad claiming in these cases somehow follows naturally from the characteristics of pioneer patents since in newly opened fields there is usually little prior art to prevent inventors from over claiming.[20] For instance, under this view, wide scopes of protection should be granted for pioneer patents over human gene inventions.

---

[15] Anna B Laakmann, 'An Explicit Policy Lever For Patent Scope' (2012) 19 Mich Telecomm & Tech L Rev 43.

[16] Robert P Merges and Richard R Nelson, 'Market Structure and Technical Advance: The Role of Patent Scope Decisions' in Thomas M Jorde and David J Teece (ed), *Antitrust, Innovation and Competitiveness* (Oxford University Press 1992; Laakmann (n 15).

[17] Tim Roberts, 'Broad Claims for Biotechnological Inventions' (1995) 17 EIPR 267.

[18] Dan L Burk and Mark A Lemley, 'Policy Levers in Patent Law' (2003) 89 Virginia Law Review, 1575.

[19] Oliver Mills, *Biotechnological Inventions, Moral Restraints & Patent Law* (Revised Edition, Ashgate 2010) 86. In contrast, Merges and Nelson argue that the granting of broad scopes of protection to important innovations (and thus limiting competition) has not resulted in an increased rate of innovation in the corresponding fields, see Robert P Merges and Richard R Nelson, 'On Limiting or Encouraging Rivalry in Technical Progress: The Effect of Patent Scope Decisions' (1994) 25 Journal of Economic Behavior & Organization 1.

[20] Burk and Lemley 'Policy Levers in Patent Law' (n 18).

Nevertheless, while the grant of broader patents may seem appropriate for inventions that represent a significant step in technical development, they are not suitable when the invention represents a minor technological advance in a no longer groundbreaking technology.[21] However, the problem is that most cases fall somewhere in between these two extremes.[22] Furthermore, given the highly innovative character of many inventions it is usually very difficult to foresee what future developments will follow or how difficult it will be to achieve them.[23] For example, what might be pioneering now may not be so innovative in a few years. Thus, in the field of genetics, where many inventions claim pioneer status, it is questionable whether all of them are entitled to it. In this regard, granting broad scopes of protection could be acceptable in the case of truly groundbreaking inventions but it becomes more difficult to justify as (formerly revolutionary) industries advance. Thus, inventions claiming pioneer status and thus broad scope of protection should be carefully examined in order to ensure the fair exchange objectives of the patent system.

Nonetheless, since broad patent scope means that more improvements will fall within the exclusive right of the patentee, and thus be excluded from the market, patent applicants are generally keen to file very broad patent applications, whether pioneer or not, in order to obtain patent protection that is as broad as possible.[24] Broad patents allow companies to capture more royalties from competing companies who seek licenses to use the patented inventions.[25] Therefore, what clients will expect from their attorneys is a broad and dominant approach to claim drafting that,[26] if granted, will allow the patent holder to control a larger part of a technical area.

---

[21] Amanda Warren-Jones, *Patenting rDNA: Human and Animal Biotechnology in the United Kingdom and Europe* (Lawtext Pub 2001) 103.

[22] Cantor (n 12).

[23] Ibid; Merges and Nelson, 'On The Complex Economics of Patent Scope' (n 6); Warren-Jones (n 21) 103.

[24] See Bostyn, 'A European Perspective on the Ideal Scope of Protection and the Disclosure Requirement for Biotechnological Inventions in a Harmonized Patent System: The Quest for the Holy Grail' (n 12); Matthew Fisher, *Fundamentals of Patent Law: Interpretation and Scope of Protection* (Hart Pub 2007) 97.

[25] Andrea D Brashear, 'Evolving Biotechnology Patent Laws in the United States and Europe: Are They Inhibiting Disease Research' (2001) 12 Ind Int'l & Comp L Rev 183.

[26] R Stephen Crespi, 'Biotechnology, Broad Claims and the EPC' (1995) 17 EIPR 267.

Another important factor influencing broad scope claims is the first-to-file system existing in Europe in which the first person that files a patent application is entitled to it if granted.[27] This race to the patent office has its advantages such as simplicity and expediency as it is simpler to determine priority of application than priority of invention.[28] However, it may also encourage premature filing of patent applications before a specific use for the invention has been identified, and thus claiming protection broader than the invention's known contribution to the art, in order to avoid losing the patent to competitors with an earlier priority date.

The issue of broad claims and optimal patent scope is especially controversial in the field of human genetics, where many inventions fall within the grey area between truly pioneer innovations and follow-on improvements. In this regard, the experience of the past three decades has led to a growing interest in establishing clearer and narrower limits to the patent scope of inventions in this field.

6.2.3 An Increasing Desire to Narrow the Scope of Biotechnological Patents

With the rapid advances in genetic sciences, a growing number of very broad claims began to be filed and sometimes granted by patent offices worldwide. For example the EPO has allowed generic claims to genetic inventions on several occasions,[29] which grant protection over a vast range of unknown advances.[30] In these cases many of the applicants' claims have been directed to research results based on a relatively modest and incomplete understanding of the functions, but patents have been granted, for example, over DNA sequences per se.[31] Furthermore, claims are often construed very broadly without clearly specifying the limits of the invention, for instance by using descriptions loaded with general and

---

[27] Convention on the Grant of European Patents (European Patent Convention (EPC)) of 5 October 1973, art 60(2).

[28] See Fisher (n 24) 98.

[29] See Sibley (n 9) 123; Roberts (n 17).

[30] See Warren-Jones (n 21) 103.

[31] See Joseph Straus, 'Patenting of Human Genes and Living Organisms – The Legal Situation in Europe' in Friedrich Vogel and Reinhard Grunwald (eds), *Patenting of Human Genes and Living Organisms* (Springer 1994) 21; Crespi, 'Biotechnology, Broad Claims and the EPC' (n 26).

irrelevant information.[32] In this sense, applications often contain open transition terms (e.g. comprising) instead of closed transition terms (e.g. consisting of).[33] By contrast, many claims also include functional terminology to define the invention by what it does (e.g. DNA sequence X that, when used in a host, is capable of expressing protein Y),[34] or that are accompanied by limited exemplification as in the Harvard Oncomouse patent case. It seems therefore that patents within the field of genetics are often designed to cover future unknown applications rather than to disclose the actual content of the invention.

Furthermore, patent offices and courts, which are the two bodies responsible of making patent scope decisions at pre-grant and post-grant stage respectively, generally have considerable room for discretion.[35] Moreover, it is virtually impossible for any institution to make sensible predictions about:

- the applications that will emerge for the patentee's invention
- the variations of the invention that might develop
- the competing or substitute technologies that will arise or
- the dependence or independence of complementary technologies to the given invention.[36]

In this regard, the flexibility of courts and patent offices and the high level of unpredictability in genetics can also result in the grant of overly broad patents in this field.

Genetics is a relatively new science where a lot remains to be discovered, and although some authors have found that broadening the scope of patent protection is associated with higher valuations (e.g. Kitch

---

[32] Tine Sommer, *Can Law Make Life (too) Simple?: From Gene Patents to the Patenting of Environmentally Sound Technologies* (1st edn, DJØF Publishing 2013) 530.

[33] Jan B Krauss and Toshiko Takenaka, 'A Special Rule for Compound Protection for DNA-Sequences – Impact of the ECJ "Monsanto" Decision on Patent Practice' (2001) 93 JPTOS 189.

[34] Bostyn, *Enabling Biotechnological Inventions in Europe and the United States: A Study of the Patentability of Proteins and DNA Sequences with Special Emphasis on the Disclosure Requirement* (n 11) 147.

[35] Merges and Nelson, 'On The Complex Economics of Patent Scope' (n 6).

[36] Dan L Burk and Mark A Lemley, 'Fence Posts or Sign Posts: Rethinking Patent Claim Construction' (2009) 157 U Pa L Review 1743.

and Lerner),[37] overly broad gene patents upset the balance that the patent system aims to strike between the inventor and the public,[38] while having a negative impact on innovation and development. Moreover, initial pioneer gene technology techniques are increasingly considered to be more predictable and are more likely to fall into the category of routine experimentation. One example of a type of technology where this has occurred is the production of monoclonal antibodies, where the nature of the technology involves a screening process.[39] In this sense, in order to control the scope of gene patents and their impact on subsequent innovation, Article 9 of the Biotech Directive restricts the scope of protection of these patents to the invention's ability to perform its function (practical contribution), which in the case of human gene patents is revealed through the requirement of industrial application set out in Article 5(3) of the Biotech Directive. This provision adds to Article 69 of the EPC, which deals with the scope of patent protection in general.

6.2.4 Economic Theories on Patent Scope

Very different theories regarding the optimal scope of patent rights coexist in the existing economic literature on patent theory. Studying the contrasting perspectives that these theories offer and their limitations helps to explain the importance of patent scope and the effect that different scope sizes might have on the innovation process in different technical fields. Besides, exploring these theories is also important for comprehending the rationale behind the current European provisions on the determination of scope of protection in general and with regard to genetic inventions.

Since an extensive review of these theories can be found elsewhere,[40] this chapter will focus on those issues that are especially relevant to the interpretation of patent scope with regard to genetic material. In essence, this part of the chapter aims to show:

---

[37] Edmund W Kitch, 'The Nature and Function of the Patent System' (1977) 20 Journal of Law & Economics 265); Joshua Lerner, 'The Importance of Patent Scope: An Empirical Analysis' (1994) 25 The RAND Journal of Economics 319.

[38] Roberts (n 17).

[39] Cantor (n 12).

[40] See for example Bostyn, *Enabling Biotechnological Inventions in Europe and the United States: A Study of the Patentability of Proteins and DNA Sequences with Special Emphasis on the Disclosure Requirement* (n 11); Fisher (n 24) 98.

- the differences between the reasoning behind each of these theories
- how these theories would apply within the context of gene patents and
- how those theories advocating for establishing a narrow patent scope, that would not deter subsequent innovation, rather than an overly broad scope of protection are supported by the Biotech Directive's provisions limiting the scope of protection of patents concerning genetic material to the functionality of the claimed invention.

6.2.4.1 Length and breadth theories

The models developed by Klemperer and Gilbert and Shapiro studied the issue of optimal patent size as a trade-off between patent length and breadth. These authors consider patent life and scope to be similar in terms of their incentive effect on inventors. Klemperer focuses on designing patents with a specific combination of length and breadth that reward innovators with a minimum social cost. His theory argues that infinitely lived but narrow patents are typically desirable when substitution costs between varieties of the product are similar across consumers, while very short-lived but wide patents are desirable when valuations of the preferred variety relative to not buying the product at all are similar across consumers.[41] According to Klemperer, a narrow patent is desirable when it causes relatively few consumers to substitute, that is, when demand is relatively inelastic in substitution cost.[42] These results suggest that optimal patent policies should vary across different classes of products[43] depending on the elasticity of demand of a certain good and the level of substitution costs.

Klemperer's analysis focuses on the link between broad scope, competing varieties of the patented product and the cost associated with substitution of the patented product. In contrast, Gilbert and Shapiro emphasize the extent to which a patentee may exploit the patent right for a given coverage of the patent grant. In their model patent breadth has no effect on the set of substitute products that are offered to consumers but can only affect the price that the patentee can charge.[44] Gilbert and

---

[41] Paul Klemperer, 'How Broad Should the Scope of Patent Protection Be?' (1990) 21 The RAND Journal of Economics 113.

[42] Ibid.

[43] Ibid.

[44] Gilbert and Shapiro (n 1). In contrast, Gallini proposes an optimal policy consisting of broad patents (no imitation allowed) with patent lives adjusted to achieve the desired reward. The key in this model is the increasing cost of

Shapiro find that the optimal length may be infinite in all cases and that the appropriate margin on which patent policy should operate would be patent breadth.[45] They interpret patent scope as the ability of the patentee to raise the price of the product covered by the patent. Within this context, and given the overall level of rewards for innovators, the authors suggest longer patent lives combined with more careful antitrust treatment of patent practices, such as the provisions of licensing contracts, so long as patent breadth is increasingly costly in terms of deadweight loss.[46] Under this model, awarding infinite patents and controlling the costs generated by broad scope patents through external intervention could achieve optimal patent design.

Although these theories contain valuable proposals regarding optimal patent design, they also present important limitations within the context of genetic inventions. The main shortcoming of both models is that they do not take into account the effect of scope size on subsequent research, an aspect that is especially important in the field of genetics where innovations build upon each other and access to patented technologies is essential. As Klemperer himself recognizes, a wider patent may prevent duplicative or imitative efforts but a narrower patent may provide greater incentives to refine and improve the original innovation.[47] On the other hand, neither does Gilbert and Shapiro's theory consider the dangers of infinitively long patents on future research. Moreover, both papers ignore the role of subsequent inventions that substitute for the original invention but also improve on it, and may result in something significantly better than the patented technology.[48] Consequently, since both models disregard the impact of patent design on follow-on innovation efforts, these theories as they stand do not provide an appropriate model for patent scope that promotes subsequent innovation in genetic research.

6.2.4.2 Patent breadth theories

In parallel to length and breadth theories, several models that focused exclusively on patent scope have also been developed. In this regard, the theories proposed by Kitch and Merges and Nelson are especially relevant given the opposite rationales behind these theories and the different effects that their ideas would have on gene patenting.

---

imitation as patent life increases, see Nancy T Gallini, 'Patent Policy and Costly Imitation' [1992] The RAND Journal of Economics 52.

[45] Gilbert and Shapiro (n 1).

[46] Gilbert and Shapiro (n 1).

[47] Klemperer (n 41).

[48] Merges and Nelson, 'On The Complex Economics of Patent Scope' (n 6).

Kitch's prospect theory supports broad patent scope to incite pioneer investors in undeveloped fields of technology to invest in and develop such fields. Kitch's theory centres on the notion that a patent gives its owner the exclusive right to further develop the 'prospect' that the invention represents to induce 'prospecting', so that the pioneer inventor will organize the market to develop improvements and refinements of the invention efficiently.[49] According to Kitch, monopoly rights present more advantages for technical advance than a rivalrous setting.[50]

(1) This exclusive right allows breathing room for the inventor to invest in development without the fear that another firm will preempt or steal his work[51] and
(2) It permits the inventor to coordinate activities with those of potential imitators, for example through licensing contracts,[52] to reduce inefficient duplication of inventive efforts and thus avoid the inefficiencies associated with rival uncoordinated innovation.[53]

In sum, Kitch's model views invention as the consequence of the efforts of a pioneer firm that should give the patentee the right to control subsequent research in that area during the whole life of the patent.

Kitch's prospect theory has found support in the writings of other authors. For example, Kitch agrees with Schumpeter to the extent that the monopoly holder should be allowed to exercise the subject of his rights completely and without interference.[54] Likewise, Chang proposes broad patents for discoveries with both extremely high social value and with low social value but related to improving technical areas.[55] Denicolo recognizes that reducing patent scope leads to more competition in the

---

[49] In his theory, Kitch analogizes patents to exclusive claims to the minerals that may be produced from a piece of land, see Kitch (n 37).

[50] Merges and Nelson, 'Market Structure and Technical Advance: The Role of Patent Scope Decisions' (n 16).

[51] Kitch (n 37). See also Merges and Nelson, 'On The Complex Economics of Patent Scope' (n 6); Burk and Lemley, 'Policy Levers in Patent Law' (n 18).

[52] Scotchmer, *Innovation and Incentives* (n 4) 152.

[53] Kitch (n 37). See also Merges and Nelson, 'On The Complex Economics of Patent Scope' (n 6); Burk and Lemley 'Policy Levers in Patent Law' (n 18).

[54] Arti K Rai, 'Fostering Cumulative Innovation in the Biopharmaceutical Industry: The Role of Patents and Antitrust' (2001) 16 Berkeley Tech LJ 813. See also Natasha N Aljalian, 'The Role of Patent Scope in Biopharmaceutical Patents' (2005) 11 BUJ Sci & Tech L 1.

[55] Howard F Chang, 'Patent Scope, Antitrust Policy, and Cumulative Innovation' (1995) 26 RAND Journal of Economics 34.

product market after the innovation. According to Denicolo competition is usually socially desirable but may also involve social costs derived from duplication of efforts like duplication of entry costs and inefficient production.[56] In their analysis of the optimal model of patent scope, O'Donoghue, Scotchmer and Thisse distinguish between lagging breadth, which protects against competition from products inferior to the patented products such as imitation products, and leading breadth, which protects against competition from products with higher quality;[57] to conclude that lagging breadth may not alone provide sufficient incentives for investment whereas leading breadth can extend effective patent life and stimulate research and development. Finally, in view of the shortcomings of letting innovators collect all the social value of their invention, as well as of awarding narrow protection so that second innovators will less easily infringe on the first innovator patent, Scotchmer proposes cooperative prior agreements between first (original) and second innovators, who improve existing inventions but create new products as well, which permit firms to share the costs and proceeds of research.[58]

On the other hand, the main criticism of Kitch's model is that his ideas are not in conformity with a fundamental principle in patent law, that is, that a patent applicant can only obtain patent protection for what he has invented, and not for future inventions.[59] Besides, holders of broad patents might be operating more as gatekeepers rather than coordinators since the ability to charge a royalty does not ensure more efficient development even if licences are granted.[60] Moreover, competition and rivalry are essential elements of the proper functioning of the patent system and thus patents with overly broad scope under the control of only a few entities might have a negative effect on technological progress.[61] In this sense, granting broad prospects to first inventors might

---

[56] Vincenzo Denicolo, 'Patent Races and Optimal Patent Breadth and Length' (1996) XLIV The Journal of Industrial Economics 263.

[57] Ted O'Donoghue, Suzanne Scotchmer and Jacques-François Thisse, 'Patent Breadth, Patent Life and the Pace of Technological Progress' (1998) 7 Journal of Economics & Management Strategy 3.

[58] Suzanne Scotchmer, 'Standing on the Shoulders of Giants: Cumulative Research and the Patent Law' (1991) 5 Journal of Economic Perspectives 30.

[59] See Roger L Beck, 'The Prospect Theory of the Patent System and Unproductive Competition' (1983) 5 Research in Law and Economics 193.

[60] Merges and Nelson, 'On The Complex Economics of Patent Scope' (n 6).

[61] See Bostyn, *Enabling Biotechnological Inventions in Europe and the United States: A Study of the Patentability of Proteins and DNA Sequences with Special Emphasis on the Disclosure Requirement* (n 11) 62 and 43.

also give rise to wasteful races between competitors.[62] Finally, strengthening patent rights does not always increase incentives to invent but greatly increases an improver's chances of being involved in litigation.[63] Therefore, in light of the increasing concern about the potential negative effects of overly broad patents on genetic information such as limited access to genetic discoveries, the limitations of Kitch's model make its application to patent scope decisions difficult with regard to these types of inventions.

Like Kitch, Merges and Nelson's incentive to competition theory, or race to invent theory, studies the question of how patent scope design influences the subsequent development of a technology; however, they conclude that granting broad scope patents will probably hamper the development of the technology rather than promote it. Through the use of historical studies, the authors argue that awarding broad scope patents to important inventions has slowed progress in several industries. Moreover, they find that patent scope influences progress in each industry differently depending on:

- the nature of the technology involved[64]
- the topography of technical advance in a field, particularly on how inventions are linked to each other[65]
- the extent to which rapid technical advance requires a diversity of actors and minds[66] and
- the extent to which firms license technologies to each other.[67]

Taking into account those factors, the authors propose at least four different generic industry models, namely:

- discrete invention
- cumulative technologies
- chemical technologies and finally

---

[62] Mark F Grady and Jay I Alexander, 'Patent Law and Rent Dissipation' (1992) 78 Virginia Law Review 305.

[63] Merges and Nelson, 'On The Complex Economics of Patent Scope' (n 6).

[64] Merges and Nelson, 'On Limiting or Encouraging Rivalry in Technical Progress: The Effect of Patent Scope Decisions' (n 19). See also Merges and Nelson, 'Market Structure and Technical Advance: The Role of Patent Scope Decisions' (n 16).

[65] Merges and Nelson, 'On Limiting or Encouraging Rivalry in Technical Progress: The Effect of Patent Scope Decisions' (n 19).

[66] Ibid. See also Ko (n 2).

[67] Merges and Nelson, 'On The Complex Economics of Patent Scope' (n 6).

- science-based technologies like genetic engineering where technical advance is driven by developments in science outside the industry.[68]

In each of these models patent scope issues take a special form.[69] Moreover, these categories are dynamic, so that a particular field of invention may fall under a different classification as the field matures.[70] Thus, the optimal patent scope in each category of industry changes as the industry advances. In the case of science-based technologies research and development efforts attempt to exploit recent scientific developments, which tend to narrow the perceived range of technological opportunities and concentrate the attention of inventors on the same projects.[71] This is a context that encourages inventive races to develop a major advance over prior art,[72] particularly in the European first-to-file system. In this regard, Merges and Nelson consider that the efficiency gains from the pioneer's ability to coordinate are likely to be outweighed by the loss of competition for improvements to the basic invention.[73] Their model thus favours a competitive environment for improvements, rather than coordinated development by the pioneer firm.

Merges and Nelson's incentive to competition theory relies on proposals like the theory of competitive innovation developed by Arrow,[74] and on evidence such as that provided by Taylor and Silberston.[75] Nonetheless, the main limitation of the theory proposed by Merges and Nelson is that, as they note, their analysis is only applicable to the broader claims

[68] Merges and Nelson, 'Market Structure and Technical Advance: The Role of Patent Scope Decisions' (n 16).

[69] Merges and Nelson, 'On The Complex Economics of Patent Scope' (n 6).

[70] Fisher (n 24) 152.

[71] Merges and Nelson, 'On The Complex Economics of Patent Scope' (n 6).

[72] Ibid.

[73] Ibid.

[74] Arrow argues against Kitch's prospect theory by asserting that competition, not monopoly, best spurs innovation. In his view, patent rights should be narrowly confined to specific embodiments of an invention and should not give the patentee monopoly control over product markets, see Kenneth J Arrow, 'Economic Welfare and the Allocation of Resources for Inventions' in Richard R Nelson (ed), *The Rate and Direction of Inventive Activity: Economic and Social Factors* (Princeton University Press 1962). See also Laakmann (n 15).

[75] Taylor and Silberston show that the patent system is regarded as essential by firms in only a small number of industries, so a reduction in the scope of protection will not severely undercut the incentive to invent, see Christopher T Taylor and Z Aubrey Silberston, *The Economic Impact of the Patent System: A Study of the British Experience* (Cambridge University Press 1973) 365.

of a small number of patents, primarily those on pioneering breakthroughs, and must always be interpreted in light of the nature of technical advances in each particular industry. Furthermore, Merges and Nelson acknowledge that competition may at times be inefficient, wasteful and duplicative.[76] Besides, a system that discriminates against pioneer inventions and promotes incremental advancement may promote secrecy, and thus more rent dissipation through duplicative efforts.[77] Moreover, the narrowness of the grant and the consequent incremental nature of technological advance might eventually promote the emergence of clusters of narrow patents that impose barriers to further research.[78] However, notwithstanding existing criticisms against this model, the incentive to invent theory responds better to concerns about the increasing number of overly broad patents in genetic industries. This theory acknowledges the importance of the specific characteristics of each technological field for designing patent scope and the role of competition in original and follow-on innovation; an approach that is consistent with the Biotech Directive's aim of avoiding granting overly broad scopes for human gene patents that extend beyond the invention's technical contribution, and thus may undermine the progress of further research.

Nonetheless, the models described above reveal an important mismatch between patent scope theories and the real world. The lack of comprehensive data on the economic effects of patents, which often leads authors to rely on anecdotal evidence to support their positions, makes it difficult to test the different patent theories.[79] Moreover, the parameters used often assume almost perfect conditions that do not occur.[80] From a theoretical point of view, the models described above offer valuable guidance about the different possible designs of patent scope. However, the key to understanding the wide range of theories for optimizing patent rules is the different industry contexts in which patents exist.[81] In this regard, the particularities of each industrial field may require the adoption of specific measures for dealing with patent scope issues in certain technologies, which impedes the unconditional application of one of those theories. Following this view, in Europe the Biotech Directive has

---

[76] Aljalian (n 54).

[77] Fisher (n 24) 154.

[78] Ibid 155.

[79] See Laakmann (n 15).

[80] Bostyn, *Enabling Biotechnological Inventions in Europe and the United States: A Study of the Patentability of Proteins and DNA Sequences with Special Emphasis on the Disclosure Requirement* (n 11) 64.

[81] Burk and Lemley, 'Policy Levers in Patent Law' (n 18).

introduced specific provisions dealing with the patent scope of genetic patents. Such provisions are additional to the EPC articles on patent scope.

6.3 PATENT SCOPE IN THE EUROPEAN PATENT SYSTEM

6.3.1 Scope of Protection under the EPC: Article 69 Background and Interpretation

A patent confers a temporary monopoly in exchange for the public disclosure of the invention to ensure widespread diffusion of its benefits to society. Following this logic the amount of protection granted should be proportional to the invention's technical contribution disclosed by the applicant. As stated by the TBA at the EPO in Exxon, there is a 'general legal principle that the extent of the patent monopoly, as defined by the claims, should correspond to the technical contribution to the art in order for it to be supported or justified'.[82] Therefore, determining the invention's contribution to the art should be a preliminary question in all patent scope decisions.

Such a contribution to the art is delineated by the disclosure requirements, which require the applicant to describe the content and borders of his invention, namely, the invention's breadth. Although scope of protection and breadth of invention are not identical, with breadth of invention determined at the pre-granting stage or in revocation proceedings after the grant, while scope of protection can only be determined after the grant once the breadth of invention is defined,[83] there is a strong link between them since what the applicant reveals is what provides the foundations for designing the scope of the patent.

The patent claims and their descriptions establish how wide the scope of protection should be. In this regard, the disclosure requirements are essential for preventing excessive claiming in terms of the range of products and processes that fall or can fall within the protection granted.[84] That is, claims drawn broadly should entail the obligation to

---

[82] *Fuel oils/EXXON* (T 0409/91) [1993] (EPO (TBA)) point 3.3.

[83] Bostyn, *Enabling Biotechnological Inventions in Europe and the United States: A Study of the Patentability of Proteins and DNA Sequences with Special Emphasis on the Disclosure Requirement* (n 11) 145.

[84] Ibid 146–147; Warren-Jones (n 21) 104.

make a correspondingly wide disclosure,[85] in order to ensure a correct deduction of the scope of the patent. Indeed, the combination of claim breadth, claim interpretation and patent scope is what determine the amount of protection awarded to European patents.

Article 69 of the EPC lays down the extent of protection of European patents. The introduction of this provision was primarily concerned with reconciling the traditional diametric opposition between the UK and Germany regarding claim interpretation. On the one hand, in countries like the UK the claims formed the periphery of the monopoly (fence posts)[86] and patentees were required to mark out in advance the whole range of applications of the inventive concept to which those rights would extend.[87] This model represented a narrow approach to interpreting claim language that only allows for literal or textural infringement.[88] By the time Article 69 was being negotiated, the UK was consistently treating claims as the literal borders of the applicant's invention, which was seen as an approach that was too narrow. On the other hand, in Germany the claims only served as guides to define the core of the invention and the scope of protection was determined by generalization of the inventive concept, even if the teaching's scope was wider than the literal reading of the patent claims.[89] The German model influenced

---

[85] Li Westerlund, *Biotech Patents: Equivalence and Exclusions under European and U.S. Patent Law* (Kluwer Law International 2002) 84.

[86] David L Cohen, 'Article 69 and European Patent Integration' (1997) 92 Nw U L Rev 1082.

[87] William R Cornish, 'Scope and Interpretation of Patent Claims under Article 69 of the European Patent Convention' (2000) 4 Int'l Intell Prop L & Pol'y 34.

[88] In fact, historically the UK courts had only departed from the literal interpretation of claims in cases where the court considered that the accused infringer had appropriated the core of the invention. However, this reluctance to broaden claims beyond their strict literal meaning was changed by the 1982 landmark decision in *Catnic Components Ltd v Hill & Smith Ltd*. From this case emerged Lord Diplock's test for infringement according to which the patent claim is given a purposive construction as opposed to a literal interpretation, see *Catnic Components Ltd v Hill & Smith Ltd* [1982] RPC 183 (HL). See also Allan M Soobert, 'Analyzing Infringement by Equivalents: A Proposal to Focus the Scope of International Patent Protection' (1996) 22 Rutgers Computer & Tech L J 189; Alexandra K Pechhold, 'The Evolution of the Doctrine of Equivalents in the United States, United Kingdom, and Germany' (2005) 87 JPTOS 411. After Catnic, the purposive construction principle was articulated and elaborated upon in *Improver Corp v Remington Consumer Products Ltd* [1990] FSR 181 (Pat).

[89] The above-mentioned principle of the interpretation of claims was used in *Absetzvorrichtung* (BGH) [1983] GRUR 497 and *Bierklärmittel* (BGH) [1984]

several smaller states, including the Netherlands, Austria and Switzerland.[90] In sum, the important part of the claim was the content (inhalt) and not the specific terms employed by the applicant.

In view of the existing opposite approaches to claim interpretation, the delegates taking part in the elaboration of a European rule regarding scope of protection attempted to create a compromise and opted for an intermediate solution between the UK and German approaches, which culminated in the adoption of Article 69 and its Protocol of Interpretation.[91] The first preparatory works for a European rule on the extent of protection were based on a study commissioned by the Expert Committee on the fundamental conditions for patent protection in the Member Countries of the Council of Europe.[92] Based on that study, a final report dated 11 October 1960 suggested that clarification of the scope of protection of the patent could be achieved if the patent claims were to be regarded as decisive for the extent of patent protection.[93] This would extend protection to all embodiments that used the essential part of the inventive notion expressed in the patent claim, even if they did not correspond with the wording of the patent claim in every detail.[94] However, the extent of protection was not to include solutions that were merely disclosed in the description or the drawings but not in the claim, although anything in the description or the drawings linked to any part of the patent claim could be used to interpret (clarify) the patent claim.[95]

This intermediate solution was included in the text proposal[96] for a first working draft of an agreement on a European patent law dated 14 March 1961, and was aimed at achieving a compromise solution for

GRUR 425 of the Federal Court of Justice of Germany, see Sasa Bavec, 'Scope of Protection: Comparison of German and English Courts' Case Law' (2004) 8 Marq Intell Prop L Rev 255. See also Cohen (n 86); Cornish (n 87).

[90] Cornish (n 87).

[91] Jochen Pagenberg and Uta Köster, 'History of Article 69 EPC' in Jochen Pagenberg and William R Cornish (eds), *Interpretation of Patents in Europe: Application of Article 69 EPC* (Heymanns 2006) 6.

[92] Ibid 2.

[93] Ibid 2–3.

[94] Ibid.

[95] Ibid.

[96] Committee of Experts, *Report of the meeting held in Paris on 16 and 17 March 1961*, EXP/Brev B (61) 3 (24 March 1961), Annexe, Art 4, page 8, see Pagenberg and Köster (n 91) 3.

the determination of a substantive extent of protection.[97] The text provisionally adopted by the Patents Working Party in a document in May 1961 provided that the extent of the protection conferred by the European patent shall be determined by the terms of the claims and notwithstanding, the description and drawings shall be used to define the extent of the claims.[98] Following the hearing at the Fifth Session of the Inter-Governmental Conference for the setting up of a European System for the Grant of Patents at the beginning of 1972, it was decided not to change the wording adopted in Article 8 of the 1963 Strasbourg Convention and to incorporate the same formulation in Article 69 of the EPC in 1973.[99] Moreover, to facilitate a uniform interpretation of Article 69 by national courts, a Protocol on the Interpretation of Article 69 (the Protocol)[100] was also included in the Convention.

In essence, the drafters of the Convention opted to follow the English approach of peripheral claim interpretation where the claims define the boundaries (and thus the scope) of the patent rather than the German central inventive concept approach. However, Article 69 of the EPC also provides that the drawings and description shall be used to interpret the claims, thus allowing for an interpretation of the claims that goes further than their literal meaning. In fact, this formulation seeks to lay down a principle for interpreting claims which is somewhere between the system where the claims may be interpreted strictly according to the letter and one where the claims wording does not play a decisive part in defining the limits of protection.[101] Article 69 was thus aimed at establishing a middle approach that included what the delegates thought were the best features of the UK and German approaches.

---

[97] *First Preliminary Draft of a European Patent Law Convention* (14 March 1961) Chairman's proposals, Articles 11–29, page 13a, see Pagenberg and Köster (n 91) 4.

[98] EXP/Brev/Misc (61) 7 (2 May 1961); see also Pagenberg and Köster (n 91) 5.

[99] Pagenberg and Köster (n 91) 8. See also Edward Armitage, 'Interpretation of European Patents (Art 69 EPC and the Protocol on the Interpretation)' (1983) 14 IIC 811.

[100] Protocol on the Interpretation of Article 69 of the Convention, adopted at the Munich Diplomatic Conference for the Creation of a European System for the Grant of Patents on 5 October 1973. The explanation in the form of an interpretation rule in the text valid today was added in the unpublished *Second Preliminary Draft of a Convention in a European System for the Grant of Patents* (March 1972), see Pagenberg and Köster (n 91) 8.

[101] Fisher (n 24) 225.

As stated above, in order to avoid conflicting interpretations of Article 69 of the EPC, a Protocol for the interpretation of this article was also adopted. Like Article 69, the Protocol suggests a halfway position between the UK and German models. It stipulates the permissible readings of the patent claims.[102] According to its text, the claims may be interpreted in a purposive manner based on their strict literal meaning, and then their scope of protection may be supplemented by consideration of equivalents that are obvious at the date of infringement of the patent.[103] Basically, all the elements claimed by the applicant need to have sufficient bases in the terms of the claims, although their interpretation is not strictly confined to the literal wording of the claims. Article 69 is to be interpreted as defining a position between a purely literal interpretation of claims and considering the wording of the claims as merely guidance. This is in order to provide a fair protection for the patent proprietor with a reasonable degree of legal certainty for third parties. Nonetheless, the protection conferred by a patent can be extended to any element that is sufficiently equivalent to an element specified in the claims.

In order to set the basis for achieving harmonization in the interpretation of patent scope among the national courts of EPC countries,[104] provisions similar to Article 69 EPC and the Protocol have been included in the national patent laws of the EPC Member States. However, cases

---

[102] Brad Sherman, 'Patent Claim Interpretation: The Impact of the Protocol on Interpretation' (1991) 54 MLR 499.

[103] Article 2 of the Protocol extends protection beyond the applicant's actual claims. This provision shares similarities with the US 'doctrine of equivalents' according to which a product that does not fall under the literal wording of the patentee's claims can infringe the patent if it is functionally or technically equivalent to the claims, see Fisher (n 24) 390; Simon Cohen and others, 'Litigating Biotech Patents in Europe' (2008) 28 IAM Magazine 45. However, the absence of a binding definition of equivalents poses difficulties to efforts for achieving a uniform interpretation of European patent claims by the national courts, see Stacey J Farmer and Martin Grund, 'Revision of the European Patent Convention & (and) Potential Impact on European Patent Practice' (2008) 36 AIPLA Quarterly Journal 419. In this context, like the language of Article 69, the theory of equivalents allows courts a large amount of leeway and discretion for applying different levels of literalism in their interpretations, see Marco Tom Connor, 'European Patents: What's New in 2008 for Applicants' (2008) 90 JPTOS 587.

[104] Bavec (n 89).

like Epilady,[105] in which the UK and German courts arrived at opposite conclusions after a lengthy appeal process, and Pozzoli,[106] where the UK and French courts revoked a patent that had been upheld in Germany and that differed in their interpretations of the claims under dispute, suggest that complete uniformity between the decisions of national courts has not been reached. Discussing these cases in depth falls outside the scope of this book, but they nevertheless show the difficulties in achieving a uniform interpretation of patent scope across Europe.

6.3.2 The Role and Interpretation of the Disclosure Requirements in Gene Patenting

As stated above, it is not possible to determine the scope of patent protection without first knowing the breadth of the invention, which is determined through the disclosure requirements of sufficiency (enabling) of disclosure and written description contained in Articles 83 and 84 of the EPC.

The disclosure requirements in patent law serve two purposes.

(1) They give public notice of the limits of the patent in order to allow society to enjoy the benefits of the invention, and permit third parties to foresee the content of the patent so that they are able to improve on and invent around it without infringment.[107]
(2) They ensure that the inventor is truly in possession of the invention at the date of filing the application,[108] that is, that the invention is reproducible and a specific industrial application is known.[109]

This second function relates to the practical utility of the invention, a factor that is essential in order to identify the invention's actual technical contribution to the art.

In the field of genetics, it is common to find cases where applicants try to get protection for products not disclosed in the application. However, notwithstanding the courts' discretion regarding how much information applicants ought to provide to the person skilled in the art,[110] claims

[105] *Improver Corp v Remington Consumer Products Ltd* [1990] FSR 181 (Pat); *Epilady VIII* (OLG Düsseldorf) [1993] GRUR Int 242.
[106] *Pozzoli v BDMO* [2007] EWCA Civ 588, [2007] FSR 37.
[107] Cantor (n 12); Westerlund (n 85) 62 and 63.
[108] Cantor (n 12).
[109] Mills (n 19) 85.
[110] Burk and Lemley 'Policy Levers in Patent Law' (n 18).

should be bound to a significant degree by what the disclosure enables, over and beyond prior art.[111] Applicants should provide more than a very general disclosure that does not correspond to the scope of the invention as claimed.[112] This is because if the claims go further than the disclosure made by the applicant, the person skilled in the art will be unable to know with certainty whether the inventor is entitled to get protection for everything claimed. Neither would the invention be sufficiently disclosed to the general public.

Nonetheless, the inherent characteristics of genetic inventions usually pose difficulties for the application of disclosure rules. For example, the variability of DNA sequences or the range of functions of a single protein makes it more difficult to assess whether an invention within this field has been made sufficiently available.[113] Furthermore, in the case of groundbreaking pioneer inventions based on little prior art, patentees find it harder to explain claims in a manner that enables peers to fully practise the invention without engaging in undue experimentation.[114] These issues often come into play when assessing the disclosure of patent claims concerning genetic inventions, which makes it difficult for patent authorities to make decisions without considering the particular circumstances of the claimed invention.

With regard to the relationship between Articles 83 and 84 of the EPC, the EPO case law has shown that, at least in the field of genetics,

---

[111] Merges and Nelson, 'On Limiting or Encouraging Rivalry in Technical Progress: The Effect of Patent Scope Decisions' (n 19).

[112] Bostyn, *Enabling Biotechnological Inventions in Europe and the United States: A Study of the Patentability of Proteins and DNA Sequences with Special Emphasis on the Disclosure Requirement* (n 11) 148.

[113] Denis Schertenleib, 'The Patentability and Protection of Living Organisms in the European Union' (2004) 26 EIPR 203.

[114] Rebecca Eisenberg, 'Analyze This: A Law and Economics Agenda for the Patent System' (2000) 53 V and L Rev 2081. In this regard it must be noted that to facilitate the enabling disclosure of biotech inventions, Rule 31 of the EPC allows the deposit of biological material according to the provisions of the Budapest Treaty on the International Recognition of the Deposit of Microorganisms for the Purposes of Patent Procedure of 28 April 1977 at the filing date. With respect to protein and gene inventions, the EPO requires additional disclosure of sequence listings in electronic form conforming to the rules laid down by the decision of the EPO President for the standardized representation of nucleotide and amino acid sequences, see Decision of the President of the EPO, OJ EPO 2011, 372 and the notice from the EPO, OJ EPO 2013, 542.

although in principle both requirements are independent[115] they are closely connected and basically concern the same matter.[116] The purpose of both requirements is basically to ensure that the grant of a patent monopoly is justified by the applicant's contribution to the art. Nonetheless, their assessments could lead to different outcomes and overly broad claims could better be criticized under one requirement or the other.[117] For example in the case of an isolated protein, it would be possible to satisfy the enablement requirement by depositing the claimed substance but not the written description requirement since it requires disclosing the invention in adequately specific terms.[118] Moreover, only Article 83 can be raised as a ground of opposition after the grant of a European patent except where amendment is sought by the patentee.[119] However, in practice disputes concerning Articles 83 and 84 arise together[120] and the distinction between them seems more semantic than real, as the EPO decision in Exxon illustrates:[121]

> In the present case … the reasons why the invention defined in the claims does not meet the requirement of Art. 83 EPC are in effect the same as those that lead to their infringing Art. 84 EPC as well, namely that the invention extends to technical subject matter not made available to the person skilled in the art by the application as filed …

On the other hand, as in other fields, there are two key questions that require careful examination when interpreting the disclosure requirements within the context of genetic inventions, which are an identification of the invention's essential features and the requirement of practicability.

---

115 *Triazoles/AGREVO* (T 0939/92) [1995] (EPO (TBA)) point 3.3.1. This distinction was also clarified in *Exxon* (n 82) point 3.3.

116 *Exxon* (n 82) point 3.5, confirmed by *Detecting sequences/ZENECA* (T 0289/96) [1999] (EPO (TBA)) point 3.3.4.

117 Westerlund (n 85) 66.

118 Cantor (n 12).

119 EPC (n 27) art 100 and rule 80.

120 Denis Schertenleib, 'The Patentability and Protection of Living Organisms in the European Union' (n 113).

121 *Exxon* (n 82) point 3.5.

6.3.2.1 The invention's essential features and the EPO's 'one way rule'

For disclosure purposes, an initial step is to identify the essential features of the claimed invention.[122] The EPO has interpreted Article 84 as a requirement to 'define clearly the object of the invention, that is to say, indicate all the essential features thereof', which are all those features that are 'necessary to obtain the desired effect or, differently expressed which are necessary to solve the technical problem with which the application is concerned'.[123] Basically, essential features are those needed to enable the skilled man to carry out the invention, although variants of essential components can also receive protection.[124] In other words, only those elements of the invention that form part of the invention's central inventive content, that is, its technical contribution to the art, need to be disclosed.

In a more recent case, the EPO has elaborated on the notion of essential features by acknowledging that:

> [T]he essential technical features may also be expressed in general functional terms, if, from an objective point of view, such features cannot otherwise be defined more precisely without restricting the scope of the claim, and if these features provide instructions which are sufficiently clear for the skilled person to reduce them to practice without undue burden.[125]

Such interpretation offers applicants the possibility of formulating their claims in a more flexible manner without revealing all the information related to the essential elements of the invention.

Related to the requirement of disclosing the essential features is the commonly cited 'one way rule' by which an invention is sufficiently disclosed if at least one example of how it can be used is described. This

---

[122] Westerlund (n 85) 65.

[123] *Control Circuit/ICI* (T 0032/82) [1984] (EPO (TBA)) point 15. See also *Amendments/XEROX* CORPORATION (T 0133/85) [1987] (EPO (TBA)); *Clarity/AMPEX CORPORATION* (T 1055/92) [1994] points 4 and 5.

[124] *Polypeptide expression/GENENTECH I* (T 0292/85) [1988] (EPO (TBA)) points 3.1.2 and 3.2.2.

[125] *Modifying plant cells/MYCOGEN* (T 0694/92) [1996] (EPO (TBA)) point 4, where the claim covered a recombinant plasmid and was defined in general terms relating to structural and functional characteristics of the invention such as plasmid, bacterium and regulon. See also *Synergistic herbicides* (T 0068/85) [1986] (EPO (TBA)) point 3.3.1; *Genentech I* (n 124) point 3.1.2; Mark D Janis, 'On Courts Herding Cats: Contending with the Written Description Requirement (and Other Unruly Patent Disclosure Doctrines)' (2000) 2 Washington University Journal of Law & Policy 55.

approach was first established by the EPO decision in Genentech I.[126] In this case the patent related to a human growth hormone and the claim related to a recombinant plasmid into which artificially made DNA encoding a desired functional protein was inserted to later express the encoded protein in a recoverable form.[127] Genentech's main claim was drafted in broad terms covering the expression of the polypeptide using any modified plasmid,[128] thus claiming other expression control sequences whose functions were still unknown.[129] However, reversing the EPO ED's previous decision, the TBA held that an invention is sufficiently disclosed if at least one way that enables the skilled person to perform the invention is clearly indicated.[130] Thus, the disclosure does not need to include specific instructions as to how all possible component variants within the functional definition should be obtained,[131] but one way to perform the invention will suffice.

The same flexible approach was adopted in Biogen. The principal claims in this case related to a recombinant DNA for use in cloning a DNA sequence in bacteria, yeasts or animals cells.[132] Biogen made a claim for the DNA sequence in any host, however, at the effective date just E. coli strains were available and other hosts only became ready for use at a later stage.[133] According to the respondents in this case, the claims were drafted broadly since the description of the invention in terms of how to carry it out was much more limited in scope.[134] However, contrary to the EPO OD's decision, the TBA followed the 'at least one way' rule stated in Genentech I.

In the Board's view, it is not necessary for the purpose of Articles 83 and 100(b) of the EPC that the disclosure of a patent is adequate to enable the skilled man to carry out all conceivable ways of operating the invention embraced by the claims, but only to the necessary extent.[135] The decision relies on an open definition relating to an unknown but

---

[126] *Genentech I* (n 124).

[127] See Mills (n 19) 86.

[128] *Genentech I* (n 124) point IV of the facts.

[129] Sibley (n 9) 124; Mills (n 19) 86–87.

[130] *Genentech I* (n 124) point 3.1.5.

[131] Ibid, point 3.1.5.

[132] *Alpha-interferons/BIOGEN* (T 0301/87) [1989] (EPO (TBA)) point I of the facts.

[133] *Biogen* (n 132) point III(ii) of the facts.

[134] Ibid, point 3.1.

[135] Ibid, point 3.2. See also Martina I Schuster, 'Sufficient Disclosure in Europe: Is There a Separate Written Description Doctrine under the European Patent Convention' (2007) 76 UMKC L Rev 491.

probably finite number of human and animal interferons of the alpha-type that may differ in structure but still represent some structural similarity in view of the affinity in hybridisation tests.[136] Moreover, since the members of the class provide end products with the same biological activity, as long as this is achieved by the invention there is no necessity to provide instructions in advance on how each and every member of the class would have to be prepared.[137]

Later cases have also followed the 'one way' rule approach.[138] However, the EPO case law on disclosure has progressively moved from the very flexible reasoning established in Genentech I, according to which a single working example established sufficiency for the entire scope in all cases, to an approach in which the description has to be sufficient over the whole claimed scope.[139] For instance, in Exxon the Board stated that 'the disclosure of one way of performing the invention is only sufficient within the meaning of Article 83 EPC if it allows the person skilled in the art to perform the invention in the whole range that is claimed'.[140] However, within the context of genetic inventions this decision may be seen as a very high standard since finding one way of performance that permits carrying out all claims of an invention is only possible in a few cases.[141] Furthermore, the TBA did not clearly indicate what has to be understood as a 'whole range' but only provided that this was to be decided on a case-by-case-basis in view of the evidence available.[142] However, although the Exxon decision recognized the importance of the circumstances of each case, it falls short of ensuring that patentees actually disclose a workable solution to a specific technical problem. In the case of gene inventions, given their intrinsic characteristics, one way of reproducibility might not necessarily achieve all the claimed products and more ways of performing the claimed invention might need to be indicated.

---

[136] *Biogen* (n 132) point 4.4.

[137] Ibid.

[138] See for example *Onco-mouse/HARVARD* (T 0019/90) [1990] (EPO (TBA)) points 3.2 and 3.3.

[139] Philip W Grubb and Peter R Thomsen, *Patents for Chemicals, Pharmaceuticals, and Biotechnology: Fundamentals of Global Law, Practice, and Strategy* (Oxford University Press 2010) 289.

[140] *Exxon* (n 82) point 3.4.

[141] Crespi, 'Biotechnology, Broad Claims and the EPC' (n 26).

[142] *Exxon* (n 82) point 3.4.

6.3.2.2 The requirement of practicability

This requirement implies that the disclosure made by the applicant has to be sufficient to enable the person skilled in the art to carry out the invention without undue experimentation (burden). At the date of the application the skilled person, having read the application as a whole and in light of the common general knowledge, must be able to put the invention into practice within the whole area that is claimed without undue effort.[143] In this regard, the major problem resides in determining the line between what might be considered as an acceptable amount of trial and error, and undue experimentation.

The question of undue burden is examined from the perspective of the man skilled in the art and, in general, the EPO Boards of Appeal have set the threshold rather high in the sense that a significant amount of experimentation is allowed without being considered to be undue. For instance, in Human immune interferon-Gamma (HIF-Gamma) the invention related to an isolated DNA sequence coding for HIF-Gamma, cloning vectors, microorganisms and cell-cultures, and processes for expressing DNA encoding human immune interferon in a recombinant host cell.[144] The TBA admitted that although in the priority application the DNA sequence coding for HIF-Gamma was disclosed, reproducibility of the whole process of expressing the gene to produce the desired interferon-gamma was still a difficult, complex and time-consuming task in 1981. Nonetheless, it was finally held that the description of the DNA sequence in 1981 enabled those skilled in the art to reproduce the invention.

A similar reasoning was applied in Human tissue plasminogen activator (Human t-PA). In this case the issue was whether the disclosure of the first cloning and expression of human t-PA in E.Coli was sufficient to ensure its workability in mammalian cells.[145] Based on the evidence provided by the applicant, the EPO found that the invention was sufficiently enabled as the skilled person could be expected to prepare without inventive skill or undue burden derivatives of human t-PA by way of either amino acid deletion, substitution, insertion, addition or replacement within the framework of routine trials. In the same way, in Kirin-Amgen[146] the patent related to the manufacture of erythropoietin (Epo) by recombinant DNA techniques and even though no deposit of

---

143 *Human t-PA/GENENTECH* (T 0923/92) [1995] (EPO (TBA)) point 22.

144 *HIF-Gamma/GENENTECH* (T 0223/92) [1993] (EPO (TBA)), point 3.1.

145 *Human t-PA* (n 143) point 44(v).

146 *Erythropoietin/KIRIN-AMGEN* (T 0412/93) [1994] (EPO (TBA)) points 74 and 75. See also Schuster (n 135).

recombinant host cells was made, the TBA reasoned that Article 83 EPC only requires a deposit if others were not able to repeat the invention at all. Thus, in the event that an invention can somehow be carried out without a deposit, no deposit is required.

In view of the decisions above it seems that practicability or reproducibility can be a real problem for gene inventions since in many cases, a general principle is described but workability of such principle for the whole area claimed cannot be tested with the disclosure made by the patent applicant. The EPO has made allowances for variability in genetics and an invention will be enabled even if there is some variability in the starting material, as long as one can obtain members of the class claimed.[147] Moreover, the factor of undue experimentation is assessed in light of the facts of each particular case but a significant degree of trial and error will be allowed as long as this does not require undue inventiveness from the person skilled in the art.

In sum, the EPO's interpretation of the EPC disclosure provisions within the context of genetic patents shows a high degree of flexibility in terms of the amount of information that applicants shall disclose in order to ascertain the invention's breadth and, based on this, its corresponding scope of protection. However, the difficulties in applying general rules on claim interpretation and patent scope to inventions over genetic material pose challenges to the implementation of fundamental patent policy goals concerned with access to genetic discoveries and the advance of research in this field. Given the significant impact of broad gene patents on subsequent innovation, there is the concern that the scope of protection for these patents should not extend beyond the invention's real technical contribution. In this respect, in order to address the different difficulties for determining the appropriate scope of patent protection of biological inventions, Article 9 of the Biotech Directive introduces an additional limitation to the scope of protection of inventions concerning genetic information by restricting the scope of this type of patent to its capability of performing its intended function.

[147] Denis Schertenleib, 'The Patentability and Protection of Living Organisms in the European Union' (n 113).

6.4 SCOPE OF PROTECTION FOR GENE PATENTS: ARTICLE 9 OF THE BIOTECH DIRECTIVE

6.4.1 Background and Reasoning

Chapter II of the Biotech Directive attempts to address the difficulties and uncertainties that arise in the application of general rules on patent scope within the context of biotechnological inventions. In response to concerns regarding the issue of exhaustion of patent rights after the first sale of a product, Article 8 lays down a general principle stating that the protection conferred by a patent on a biological material shall extend to any derived biological material possessing the same characteristics as the initial biological material. This principle is further articulated in Article 9 with regard to gene inventions, which provides that the protection conferred by a patent on a product containing or consisting of genetic information shall extend to all material in which the product is incorporated and in which the genetic information is contained and performs its function. Article 9 does not intend to substitute existing rules on scope of protection but it provides additional clarification to the provisions of Article 69 of the EPC on scope of protection within the context of genetic inventions.

In view of the fact that living matter is self-replicable and may thus cause problems in respect of further generations, the 1988 First Directive Proposal aimed to address concerns regarding the issue of patent scope for living matter.[148] In this sense, it explained with regard to Article 11 (later Article 8(1)) that:[149]

> Once a patented product has been placed on the market by a patentee or with his consent, no control over the further use of the product in intra-Community trade may be exerted by the patentee or a licensee … . Article 11 will ensure that the use which is intended in a sale of patented self-reproducing material is not confused with a use which involves patent infringement. The provisions of Article 11 are needed because the issue of the extent of patent rights in respect of patented living or self-replicating material has not been dealt with in any national patent system and the provisions of the EPC do not address this question … . For national patent laws, it needs to be legislatively established that use which involves propagation solely for the purpose of obtaining additional propagative or self-replicating material does not come

---

[148] Commission of the European Communities proposal for a Council Directive on the legal protection of biotechnological inventions (1988) [COM(88) 496 final – SYN 159] OJ C10/3, 13.

[149] Ibid 47–48.

within the scope of intended use which would be exhausted upon the sale of a patented product.

Because after sale purchasers would be able to reproduce the patented material without infringement, the Commission found first sale exhaustion to be a threat to the capturing value of patents concerning easily propagating material such as seeds.[150] If the scope of protection did not extend to the material in which the protected material is incorporated, patent protection would be practically useless in some cases.[151] Therefore, in order to avoid evasion of patent infringement by simply propagating the original invention to obtain derived material with identical properties, Article 8 extends patent protection to biological material derived from the patented product or process that possesses the same characteristics as the patentee's invention.

Within this context of extended protection to cover biological material derived from and similar to the patent, Article 9 clarifies how such extension of patent protection shall be applied to gene patents. Article 13 (now Article 9) of the First Directive Proposal stated that: 'the protection for a product consisting of or containing particular genetic information as an essential characteristic of the invention shall extend to any products in which said genetic information has been incorporated and is of essential importance for its industrial applicability or utility.'[152] This approach was triggered by the broad scope of the first generation of DNA patents often covering, for example, additional sequences that hybridize with the specifically disclosed DNA or its fragments.[153]

Thus, with this provision the Biotech Directive sought to prevent patent holders from obtaining exclusive rights over products in which the genetic information is not able to perform its desired function. For example, applicants often claim several functions for the same genetic sequence but the gene either does not perform all of them, or after some time the gene function is lost and the invention is no longer capable of expressing its intended function.[154] In this regard, Article 9 would impede applicants from obtaining protection beyond the invention's ability to be used for its intended purpose. Such an approach is consistent with the

[150] Michael A Kock, 'Purpose-Bound Protection for DNA Sequences: In through the Back Door?' (2010) 5 JIPLP 495.

[151] Sven Bostyn, 'The Patentability of Genetic Information Carriers' [1999] IPQ 1.

[152] First Directive Proposal (n 148) 77 (regarding Article 13).

[153] Kock (n 150).

[154] Ibid.

utilitarian view of patents as a means to provide society with useful innovative technologies, and with Locke's theory of property as a natural right, according to which the scope of the property should be limited to the purpose behind the labour, which is directed to a useful end.

The wording of Article 13 of the First Directive Proposal remained essentially unchanged during the ten-year period of negotiations prior to the final adoption of the Biotech Directive.[155] In this regard, subsequent amendments to the initial proposal reaffirmed the Biotech Directive drafters' intention to limit the protection conferred to a genetic patent to the extent that the genetic information is contained and expressed. This requirement was included in Article 9 of the text finally agreed, which provides that in the case of inventions concerning genetic information, protection must extend to all derived materials in which the genetic information is contained and which performs its function.

6.4.2 Interpretation of Article 9: Purpose-Bound Protection versus Absolute Product Protection

Together with Article 5(3) and related recitals, Article 9 gives to the criterion of industrial applicability a critical role in the patentability of gene sequences by limiting the extension of patent protection to only those derived materials in which the genetic information carries out its function, that is, its industrial application. The link between the requirement of industrial application and Article 9 can be found in the wording of Article 13 (now Article 9) of the First Directive Proposal, which limited the protection for an invention consisting of or containing particular genetic information to those products in which said genetic information had been incorporated and is essential for its industrial applicability or utility. Although the final wording of Article 9 refers to 'function' instead of 'industrial applicability' or 'utility', it seems clear that in the legislators' view, function, industrial application and utility are equivalent concepts.

In this regard, since the function of human gene patents is the one disclosed according to Article 5(3) of the Biotech Directive, Article 9 can be seen, at least within the context of human gene inventions, as an extension to the general requirement that a patent application must

---

[155] Amended Proposal for a Council Directive on the legal protection of biotechnological inventions (1992) [COM(92) 589 final – SYN 159] OJ C44/36, 21 (Article 12 amending Article 13); Commission proposal for a European Parliament and Council Directive on the legal protection of biotechnological inventions (1996) [COM(95) 661 final] OJ C296/4, Article 11.

describe the industrial application of human gene sequences and fragments thereof in order to receive patent protection. In this regard, it follows the question of whether the relationship between Articles 5(3) and 9 of the Biotech Directive results in the establishment of purpose-bound protection for human gene patents, and thus, whether for example using the same DNA sequence for expression of a different protein or a similar sequence for expression of the same or similar protein is still included in the scope of the patent.[156] Moreover, in the case of artificially created human DNA, disclosure of the function is not required and thus absolute product protection should still be available.[157]

Besides, through this function-related requirement, Article 9 could be interpreted as introducing purpose-bound patent protection for all types of genetic inventions; that is, limited to the extent that the patented invention performs the specific function initially claimed by the patentee. Nonetheless, for other types of biotechnological inventions absolute product protection is still available under Article 3(1).[158] Notwithstanding the arguable adequacy of adopting such an approach for all sorts of genetic patents, the Biotech Directive's text and preparatory works suggest that this delimitation was sought for human gene sequences. Otherwise, it would not make sense to refer to the function or industrial application of gene patents, which establishes a connection between Article 9 of the Biotech Directive and the requirement in Article 5(3) that the function of a sequence or partial sequence of human DNA needs to be specifically disclosed in the patent application.

Furthermore, with regard to the patentability of the human body, in order to avoid any extension of the patent protection for an element isolated from the human body to the human body itself,[159] Article 9 of the Biotech Directive is applied apart from as provided in Article 5(1), which forbids the patentability of the human body and its elements.[160] This interpretation of Article 9 as a requirement of purpose-bound

---

[156] Herbert Zech and Jürgen Ensthaler, 'Stoffschutz bei gentechnischen Patenten – Rechtslage nach Erlass des Biopatentgesetzes und Auswirkung auf Chemiepatente' [2006] GRUR 529.

[157] Ibid.

[158] Sommer (n 32) 261.

[159] EC Report from the Commission to the Council and the European Parliament: 'Development and Implications of Patent Law in the Field of Biotechnology and Genetic Engineering' [COM(2002) 545 final] (First Commission Report 2002) 22.

[160] However, Article 9 of the Biotech Directive would extend the protection of a claim relating to an isolated human gene sequence to a transgenic animal in which the human gene sequence has been incorporated and is expressed, see

protection for human genetic information restricted to the prohibition of Article 5(1) is the one that has been followed by several national courts, the CJEU and the European Parliament.

Moreover, several conditions have to be met in order to apply the limited extension of protection provided by Article 9. The first requirement, that the protected genetic information is contained in the derived material, is not very difficult to fulfill since by simply incorporating the product containing or consisting of genetic information in other biological material, the latter will also contain the genetic information.[161] The second condition requiring the expression of the genetic information in the derived material is however more difficult to comply with, as was seen in the Monsanto case before the CJEU and various national courts.[162]

In this regard, together with the requirement of Article 5(3) of signalling the industrial applicability of human gene inventions, Article 9 moves closer to a more rigorous approach towards the patentability of human genetic information. Through these provisions, the technicality (practicability) requirement as it is now interpreted within the context of human gene patents helps to strike the balance between theory and application since only that knowledge that is actually being applied to a practical end will be protected from infringement.[163] It is precisely in its impact on the scope of patent monopolies that the real usefulness of the industrial applicability requirement resides.[164] Abandoning it would broaden the permissible scope of patents, thus shifting the demarcation line between appropriable and public domain knowledge.[165] For some, however, such a limitation is neither appropriate nor necessary in order to avoid extensive patenting of genetic sequences, since the preconditions for the grant of a patent over genetic discoveries would inherently restrict the scope granted.[166] Under this view, absolute product protection should be granted if the finding and structure clarification of the DNA sequence

---

Gerald Kamstra and others, *Patents on Biotechnological Inventions: The EC Directive* (Sweet & Maxwell 2003) 50.

[161] Bostyn 'The Patentability of Genetic Information Carriers' (n 151).

[162] In the case of C-428/08 *Monsanto Technology LLC v Cefetra BV and Others* [2010] ECR I-06765, Monsanto failed to demonstrate that the meal product resulting from its patented soybeans was able to perform the function (improved herbicide resistance) claimed in the patent application.

[163] William van Caenegem, 'The Technicality Requirement, Patent Scope and Patentable Subject Matter in Australia' (2002) 13 AIPJ 309.

[164] Ibid.

[165] Ibid.

[166] Zech and Ensthaler (n 156).

as such are based on an inventive step,[167] which consists of identifying and isolating a desired gene sequence before anyone else has done so.

Nevertheless, if absolute product protection were given to an isolated gene, the scope of its protection would be far-reaching since it would include not only the discovered sequence but also all future uses of such sequence in its natural form; whereas purpose-bound protection would create the balance of rights and obligations referred to in Article 7 of the TRIPS Agreement.[168] A strict requirement for purpose-bound protection ensures that access to human genetic material is still allowed for other purposes.[169] Nonetheless, restricting the scope of gene patents to the disclosed purpose while maintaining the principle of absolute product protection for all other technical fields might be considered to be an unjustified differential treatment contrary to the non-discrimination requirement of Article 27(1) of the TRIPS Agreement.[170] The compatibility between Article 9 of the Biotech Directive and the TRIPS Agreement was discussed in the CJEU's decision in Monsanto (see section 6.4.4).

Despite existing arguments in favour of absolute product protection for all types of biotechnological inventions, purpose-bound protection is widely seen as a suitable alternative for patents concerning isolated human gene sequences as such. For example, in a 2003 survey carried out by the Swiss Federal Institute of Intellectual Property, most biotech research institutions and private companies agreed that an absolute protection of DNA patents would hamper research and further development and that a concrete disclosure of the function of DNA patents would enable the restriction of patent claims.[171] With scope of protection

---

[167] Joseph Straus, 'Product Patents on Human DNA Sequences: An Obstacle for Implementing the EU Biotech Directive?' (2003) 50 Advances in Genetics 65.

[168] See TRIPS Agreement (n 3) art 7 stating that 'the protection and enforcement of intellectual property rights should contribute to the promotion of technological innovation and to the transfer and dissemination of technology, to the mutual advantage of producers and users of technological knowledge and in a manner conducive to social and economic welfare, and to a balance of rights and obligations'.

[169] Sommer (n 32) 263.

[170] Wolrad Prinz zu Waldeck und Pyrmont, 'Special Legislation for Genetic Inventions – A Violation of Article 27(1) TRIPS?' in Wolrad Prinz zu Waldeck und Pyrmont and others (eds), *Patents and Technological Progress in a Globalized World: Liber Amicorum Joseph Straus* (Springer 2009) 304.

[171] The results of the survey show that only companies with more than 250 employees were hesitant about including a limitation of the scope of protection in the patent legislation, see Swiss Federal Institute of Intellectual Property,

limited to the invention's function, investors may be more likely to contribute to the study of DNA sequences because they know that if another scientist identifies a function for a sequence, that scientist will only gain protection for that application, and that future identification of new functions will still be allowed and profits will still be available.[172] Furthermore, licensing fees would only be paid if researchers focus on the same function of the sequence.[173] Therefore, it seems that given the magnitude of the potential impact of granting absolute product protection on the development of genetic sciences, purpose-bound protection for human gene inventions presents a viable solution to address existing concerns about overly broad gene patents and ensure continuous research.

6.4.3 Interpretation of Article 9 by the Commission and the Parliament

In addition to commentators and industry stakeholders, the Commission and the Parliament have also given their views on the interpretation of Article 9 of the Biotech Directive as a requirement for purpose-bound protection.

The interest of the Commission regarding the need to clarify the interpretation of Article 9 was first stated in the 2002 Commission report submitted under Article 16(c) of the Biotech Directive.[174] Given the rapid advances in the fields of biotechnology and genetics, the Commission urged the review of two particular questions, the patentability of human embryonic (pluripotent) stem cells and cells lines, and the scope to be conferred on patents relating to sequences or partial sequences of genes isolated from the human body.[175] With regard to the second question, the Commission concluded that certain provisions of the Biotech Directive, such as those relating to the scope of protection to be granted to gene

---

'Research and Patenting in Biotechnology: A Survey in Switzerland' (Swiss Federal Institute of Intellectual Property 2003) 51 and 54–55; Nikolaus Thumm, 'Patents for Genetic Inventions: A Tool to Promote Technological Advance or a Limitation for Upstream Inventions?' (2005) 25 Technovation 1410.

[172] Erin Bryan, 'Gene Protection: How Much Is Too Much – Comparing the Scope of Patent Protection for Gene Sequences between the United States and Germany' (2009) 9 J High Tech L 52.

[173] Ibid.

[174] First Commission Report (n 159).

[175] Ibid 28.

sequences, appear to give Member States some flexibility in its transposal into national law, which might give rise to differing interpretations.[176] In particular, consideration should be given to the scope to be conferred on patents involving DNA sequences and proteins deriving from those sequences, as well as those based on ESTs and SNPs.[177] Essentially, the question to be reviewed was whether patents on gene sequences should be allowed according to the classical model of patent claim (absolute product protection covering possible future uses of the claimed sequence), or whether the patent should be restricted so that only the specific use disclosed in the patent application can be claimed (purpose-bound protection).[178]

In order to address the questions laid out by the 2002 report, the Commission submitted a second report in 2005. With regard to the scope of protection of human gene patents, the Commission acknowledged that after examining the Biotech Directive's provisions on patent scope, no indication was found that these articles address the concept of a restricted scope of protection relating to the specific use identified for the gene sequence concerned.[179] However, the report noted the intent of Articles 8 and 9 to extend the protection conferred by a patent to any biological material obtained from the claimed product, or in which the claimed product is incorporated, and where the same genetic information also expresses its function. In this regard, Article 9 would limit the scope of protection of genetic patents not to the function of the invention but to the functionality of it, that is, to the extent that the patented genetic information is able to work, to gene expression.[180] Nonetheless, in the case of human genes, the specific function of the claimed sequence must be disclosed in the patent application and therefore, the scope of protection of these patents would, in any case, be limited to the extent that the genetic material can perform the function described by the applicant.

The report then recognized the possibility of interpreting Article 9 as a requirement for purpose-bound protection since a reading of Article 5(3) and Recitals 23 and 25 together might suggest that the Community

---

[176] Ibid 22 and 27.

[177] Ibid 22 and 23.

[178] EC Report from the Commission to the Council and the European Parliament – Development and implications of patent law in the field of biotechnology and genetic engineering (SEC (2005) 943) of 14 July 2005 [COM(2005) 312 final] (Second Commission Report) 3.

[179] Second Commission Report (n 178) 3 and 4.

[180] Supporting this view see Kock (n 150).

legislator had intended to at least raise the possibility of a limited scope of protection covering only the specific industrial application identified in the patent, as far as this particular type of invention is concerned.[181] Otherwise, Article 5(3) requiring the industrial application of a human gene sequence to be disclosed in the patent application would merely repeat a standard requirement of general patent law.[182] It seems therefore that the Commission does not completely rule out the possibility of purpose-bound protection, at least within the context of human gene patents; and that, furthermore, it recognizes a logical link between Articles 9 and 5(3) of the Biotech Directive.

Although the meeting of the informal Group of Experts in March 2003 concluded that there were no objective reasons to create a specific regime of purpose-bound protection for gene patents, it left two questions unanswered:

(1) whether gene sequences isolated from the human body should be given different treatment to chemical substances on ethical grounds[183] and
(2) whether it is more valuable to society to allow broad scope of protection to the first inventor or whether a patent on a gene sequence should be limited in scope to allow future uses of such sequences to be patented freely.[184]

In view of the different existing approaches between Member States in their interpretations of Article 9 and the dangers of overly generous patents for subsequent innovation, the Commission opted not to take a clear standpoint on the interpretation of Article 9 as purpose-bound protection as opposed to classical absolute protection, but decided to continue monitoring the potential economic consequences of possible divergences between Member States' legislation.

Following the second Commission report, the Parliament published a resolution in 2005 attempting to answer the question of whether a patent on DNA should cover only the specific use disclosed in the patent application, or whether other functions and possible future uses should

[181] Second Commission Report (n 178) 4.
[182] Ibid 4.
[183] Ibid.
[184] Ibid.

also be covered by the patent.[185] In response to this question, the Parliament has called for a further limitation of DNA patents to the concrete application provided for them, and for the scope of the patent to be limited to that application allowing others to use or patent the same sequence for other applications, that is, purpose-bound protection.[186] This interpretation of Article 9 is also shared by the CJEU, as it made clear in the Monsanto decision and various national laws.

6.4.4 The CJEU Decision in *Monsanto v Cefetra*

In 2005, a patent case arose in several European countries concerning a genetically modified soybean plant, which resulted in the first CJEU decision interpreting Article 9 of the Biotech Directive. With this decision, the CJEU began to exercise its role in shaping EU patent policy.[187] The referral to the CJEU related to a dispute between a European patent (EP0546090 (hereinafter EP 090)) granted to Monsanto on a plant gene isolated sequence with the capacity of rendering the Roundup Ready (RR) soybean plant herbicide resistant, and two European companies, Cefetra B.V. (Cefetra) and Alfred C. Toepfer International GmbH (Toepfer). This RR soybean was cultivated on a large scale in Argentina, where there was no protection for the plant gene, and then imported to the Netherlands as soy meal by Cefetra and Toepfer.[188] In the Netherlands and other EU countries (including Denmark, Spain and the UK) the DNA sequence utilized in the RR soybeans was however covered by the Monsanto European patent EP 090.

The Monsanto patent was granted on 19 June 1996 and related to 'Glyphosate tolerant 5-enolpyruvylshikimate-3-phosphate synthases'. Glyphosate is a non-selective herbicide that blocks an enzyme responsible for the synthesis of aromatic amino acids (Class I enzyme 5-enol-pyruvylshikimate-3-phosphate synthase (EPSPS enzyme)). This enzyme plays an important role in the growth of plants and thus can cause their death when inhibited. Monsanto however found a class of EPSPS enzymes that are not sensitive to glyphosate, and thus, allow

---

[185] European Parliament Resolution on Patents for Biotechnological Inventions of 26 October 2005 [2006] OJ C 272E/440 (Parliament resolution 2005) points J and K.

[186] Ibid, point 5.

[187] Kali Murray and Esther van Zimmeren, 'Dynamic Patent Governance in Europe and the United States: The Myriad Example' (2011) 19 Cardozo J Int'l & Comp L 287.

[188] *Monsanto v Cefetra* (n 162) points 18 and 19.

plants containing such enzymes to survive the effects of glyphosate. Monsanto inserted those genes into the DNA of a soy plant (RR soybean plant), thus making it glyphosate-herbicide resistant.[189] As a result, the Monsanto RR soybean invention allows farmers to spray their fields with herbicide that kills the weeds but doesn't damage the crops.[190] This patent did not have any claims to soy meal but only claimed the DNA sequence used in the genetic modification process. In principle, Monsanto could thus only claim infringement against the derived soy meal if the conditions of Article 9 of the Biotech Directive were met.

For procedural reasons, Monsanto could not obtain a patent for this invention in Argentina, which encouraged other companies to grow large amounts of transgenic soybean containing the EPSPS gene in that country[191] without permission from Monsanto. After being grown in Argentina, the soybeans were pressed and processed to obtain the soy oil and the residual material (soy meal) was crushed, dried, heated and pressed into an end product that is used as animal feed.[192] This soy meal was then imported into the Netherlands where Monsanto's invention was covered by the EP 090. To avoid infringement, Monsanto sought to prevent imports containing the genetically modified soybean to a number of European territories,[193] including the Netherlands where in 2005 and 2006 three consignments were detained by customs authorities under Regulation 1383/2003[194] for suspected infringement of the Monsanto patent, where it was confirmed that the soy meal contained the RR soybean.[195]

---

[189] Ibid, points 15, 16 and 17. See also Geertrui van Overwalle, 'Editorial, The CJEU's Monsanto Soybean Decision and Patent Scope – As Clear As Mud?' (2011) 42 IIC.

[190] Van Overwalle (n 189).

[191] Philip Webber, 'Limitation of Gene Patents to Functioning DNA' (2011) 17 Journal of Commercial Biotechnology 201.

[192] Charles Gielen, 'Netherlands: Patents – Biotech Directive – Scope of Protection' (2010) 32 EIPR 93.

[193] Monsanto filed suits in Denmark, the UK (*Monsanto v Cargill* [2007] EWHC 2257 (Pat)) and Spain (*Monsanto Technology LLC v Sesostris SAE*, 'Roundup Ready Spain' (Juzgado de lo Mercantil (Commercial Court of Madrid)) Decision No 488/07 [2007]).

[194] Council Regulation (EC) No 1383/2003 of 22 July 2003 concerning customs action against goods suspected of infringing certain intellectual property rights and the measures to be taken against goods found to have infringed such rights.

[195] *Monsanto v Cefetra* (n 162) point 20.

In order to stop or avoid infringement of its patent, Monsanto sued Cefetra and Toepfer for importing the soy meal into the Netherlands. The two cases were joined[196] and the District Court of The Hague pronounced a first interlocutory judgment on 19 March 2008, finding that the subsequent crushing, separation and treatment stages were too drastic to still assume a direct relationship between the methods and the soy meal and thus the soy meal did not constitute a product directly obtained from the patented method.[197] Then a second judgment on 24 September 2008 submitted four preliminary questions to the CJEU under Article 267 TFEU.

6.4.4.1 Referral questions and answers by the CJEU

The main question referred to the CJEU related to the interpretation of Article 9 of the Biotech Directive. In this regard, the court asked whether a patent on a DNA sequence can protect any product containing the patented genetic information, even though the protected sequence is not performing its function of being herbicide resistant but had performed its function in the past (in the soy plant) and could also perform it in the future if isolated and inserted into a plant cell.[198] In other words, the question was whether the Biotech Directive provides protection for products deriving from the patented genetic material regardless of whether the protected gene sequence actually performs its function.

The other three questions related to the exhaustive character of Article 9 over national patent law; the retroactive effect of the Biotech Directive, that is, whether Article 9 is applicable to patents granted prior to its adoption; and whether Articles 27 and 30 of the TRIPS Agreement should influence the interpretation of Article 9.

6.4.4.1.1 The first question: does Article 9 confer purpose-bound protection? With regard to the first question, it was crucial that not RR soybeans as such, but dead soy meal, in which the genetic information can only be found in a residual state, was imported.[199] The Biotech Directive does not contain any provisions regarding dead matter since such material cannot replicate itself and thus does not need special

---

[196] *Monsanto v Cefetra BV, Cefetra Feed Service BV, Cefetra Futures BV and Alfred C Toepfer International GmbH* (First Instance Court in the Hague (Rechtbank's-Gravenhage)) Joined Cases No 249983/HA ZA 05-2885 and 270268/HA ZA 06-2576 [2008].

[197] Kock (n 150).

[198] *Monsanto v Cefetra* (n 162) point 32.

[199] Ibid, point 37.

legislation.[200] However, since the DNA sequence had performed its function in the soy plants, and could be isolated to perform its function again in an adequate host, the parties disagreed on whether the DNA was actually performing its function in the imported meal.

Following the Advocate General Mengozzi's opinion,[201] the CJEU held in paragraphs 38 and 39 of the decision that:

> [T]he protection provided for in Article 9 of the Biotech Directive is not available when the genetic information has ceased to perform the function it performed in the initial material from which the material in question is derived.
>
> It also follows that that protection cannot be relied on in relation to the material in question on the sole ground that the DNA sequence containing the genetic information could be extracted from it and perform its function in a cell of a living organism into which it has been transferred.

In interpreting the requirement of performing the claimed function of the genetic material set out in Article 9, the CJEU referred to Article 5(3) and related recitals of the Biotech Directive regarding the industrial application of human gene sequences. Especially relevant is paragraph 45 of the decision stating that '[s]ince the Directive thus makes the patentability of a DNA sequence subject to indication of the function it performs, it must be regarded as not according any protection to a patented DNA sequence which is not able to perform the specific function for which it was patented'.[202]

By referring to the Biotech Directive's provisions on industrial applicability, the CJEU made clear that the function referred to in Article 9 corresponds with the invention's utility disclosed in the patent application as required in Article 5(3). Basically, the CJEU admitted that the Biotech Directive's express requirement of disclosing a real and present function in the patent application should be read as incorporating the function of the gene into the concept of the DNA sequences as patentable inventions and, therefore, the scope of protection of such patents should be equally restricted to that function.[203] Thus, if the genetic sequence cannot

---

200 Kock (n 150).

201 C-428/08 *Monsanto Technology LLC v Cefetra BV and Others* [2010] ECR I-06765, Opinion of Advocate General Mengozzi.

202 *Monsanto v Cefetra* (n 162) point 43, 44 and 46.

203 Gareth Morgan and Lisa A Haile, 'A Shadow Falls over Gene Patents in the United States and Europe' (2010) 28 Nat Biotech 1172.

perform the function as claimed, the patent claim should be unenforceable.[204] With this decision, the CJEU discarded the possibility of interpreting Article 9 as a requirement of absolute product protection and gave his reasons for ruling purpose-bound protection for gene patents in which the scope of protection corresponds with their actual industrial application.

However, even though such interpretation of Article 9 seems logical with regard to claims over isolated human gene sequences as such, which are subject to the requirement of Article 5(3), abolishing absolute product protection for all types of genetic inventions is not consistent with the text of the Biotech Directive. In this sense, some authors have seen the CJEU's reference to Article 5(3) and related Recitals as an erroneous mix of concepts. According to this view, the requirement of disclosing the function of a gene is to satisfy the patentability requirement of industrial application, not to restrict the scope of properly granted patent claims directed at genes the function of which had been identified and disclosed in the patent application.[205] Article 5(3) requires disclosure of the function in the patent application, which suggests disclosure in the specification and no limitation in the claim.[206] It has been argued that with this decision the CJEU equates a DNA that no longer performs its function in a specific embodiment to a DNA sequence that has no function.[207] And even if one reads a requirement of purpose-bound protection from Article 5(3), that is still not the same as saying that the claimed DNA sequence must perform its function at all times.[208] Furthermore, Article 5(3) relates to human DNA, while the Monsanto invention concerns plant genetic material. In this regard, this importation by the CJEU might be read as an unjustified imposition of additional requirements to gene inventions that does not find support in the Biotech Directive.

Nevertheless, within the context of human gene patents, both Recital 23 relating to the requirement of industrial application and Article 9

---

[204] Amanda Odell-West, 'Has the Commodore Steered the Fleet onto the Rocks? Biotechnology and the Requirement for Industrial Applicability' (2013) 4 IPQ 279.

[205] Morgan and Haile (n 203).

[206] Kock (n 150); Sven JR Bostyn, 'A Decade After the Birth of the Biotech Directive: Was It Worth the Trouble?' in Emanuela Arezzo and Gustavo Ghidini (eds), *Biotechnology and Software Patent Law* (Edward Elgar 2011) 231.

[207] Bostyn, 'A Decade After the Birth of the Biotech Directive: Was It Worth the Trouble?' (n 206) 232.

[208] Ibid 233.

regarding scope of protection, refer to the 'function' of the gene, a word that is used in the Biotech Directive as being equivalent to 'industrial application'. Furthermore, if mentioning that the function is a constitutive element of a human DNA invention without disclosing the function in the patent application, there is no technical teaching; then the scope of the patent should be limited to the function disclosed.[209] A reading of Article 5(3), Recital 23 and Article 9 together suggests that it was the legislator's intention to establish purpose-bound protection for, at the least, patents on human genetic material. The CJEU and the Parliament have also taken this approach, although they have arguably correctly extended this interpretation to all types of genetic patents. In this regard, the CJEU's reference to Article 5(3) in a case concerning plant genetic material suggests that Article 5(3) could be applicable to non-human gene patents, thus extending the impact of this provision to all types of inventions concerning genetic information.

The CJEU's interpretation also hints at the decisions of the UK Supreme Court in Schütz and the German Federal Court of Justice in Palettenbehälter II with reference to the patentability requirement of inventive step,[210] which apply the same underlying logic for defining the scope of protection of genetic patents followed in Monsanto with reference to the requirement of industrial application, that is, limiting the scope of genetic patents to the function disclosed by the patentee in the patent application. In these decisions, the main factor for deciding infringement was whether the alleged infringer had remade the inventive concept/identity of the patented invention, that is, whether the inventive technical effects of the patented invention are embodied in the allegedly infringing products. Such an approach is also consistent with the European interpretation of copyright scope with reference to the copyright law requirement of originality for the subsistence of copyright protection.[211]

---

[209] Ibid 228.

[210] In the UK, see *Schütz (UK) Limited v Werit (UK) Limited* [2013] UKSC 16, [2013] RPC 16. In Germany, see *Palettenbehälter II* (Pallet Container II) (BGH) (German Federal Court of Justice) [2012] GRUR 1118, [2013] IIC 351.

[211] See *Infopaq International A/S v Danske Dagblades Forening* (C-5/08) [2009] ECR I-06569, interpreting Article 2 of Directive 2001/29 of the European Parliament and of the Council of 22 May 2001 on the harmonization of certain aspects of copyright and related rights in the information society. In Infopaq, the CJEU interpreted the exclusive right of the author provided in Article 2(a) of Directive 2001/29 as covering the reproduction of any elements of a work that are the expression of the intellectual creation of their author and thus reflect the originality of the work in question. In the UK, see *Designers Guild Ltd v Russell Williams (Textiles) Ltd (Trading as Washington DC)* [2000] 1 WLR 2416 (HL).

Thus, these interpretations of patent law seem to reject the concept of inventions as something distinct and independent from the secondary patentability requirements, but suggest that what makes a specific creation worthy of patent protection may reside in particular properties and that, consequently, the extent of the protection granted to an invention ought to be confined to the existence of such properties.

6.4.4.1.2 Second and third questions: does Article 9 preclude national law from awarding absolute product protection to gene patents? And if so, does it apply retroactively? The second question to the CJEU sought clarification with regard to the relationship between the Biotech Directive and national patent law. The question was essentially whether Article 9 provides a minimum standard for protection in addition to the general provisions of national patent law, or whether it is exhaustive and overrules national patent law so that a DNA sequence is further required to perform its function to establish infringement.[212] This question arises from the fact that national patent authorities disagree on what the obligations imposed are,[213] and thus differ in their interpretations of the Biotech Directive. In response, the CJEU referred to Recitals 3, 5, 7, 8 and 13 to conclude that Article 9 reflects the Community legislature's intention to ensure the same protection for patents in all Member States, so that the effect of this provision must be regarded as exhausted,[214] no matter whether the patent in question was filed before or after the implementation of the Biotech Directive.[215]

As the Advocate General noted, the fact that the rules are incomplete does not mean that they are not exhaustive, but the system established by that measure is exhaustive with regard to the particular matters dealt with, giving freedom to national legislature only in those areas where the EU legislature has not intervened.[216] This is because extended protection would compromise the balance sought between the interests of patent holders and other operators and, on the other hand, it would give rise to differences between the Member States, which would ultimately foster trade barriers.[217] Nonetheless, the practical effects of Article 9 are limited since a corresponding provision has not been included in the EPC;

212 Kock (n 150).
213 Krauss and Takenaka (n 33).
214 *Monsanto v Cefetra* (n 162) points 58 and 60.
215 Ibid, point 69.
216 *Monsanto v Cefetra*, Opinion of Advocate General Mengozzi (n 201) point 47.
217 *Monsanto v Cefetra* (n 162) point 59.

although European countries are progressively introducing legislation expressly limiting the scope of protection of gene patents, especially those of human origin, to the disclosed function.

For example, Articles L611-18(2) of the French Intellectual Property Code limit the protection of any invention relating to an element of the human body to the extent necessary for the realization and exploitation of its particular use; while Article L6113-2-1 restricts the scope of any gene sequence claim to the disclosed application.[218] Likewise, the interplay of paragraphs 3 and 4 of the new section 1a of the German Patent Act effectively restricts the scope of patents for human gene sequences to their disclosed purpose by requiring patent applicants to disclose and claim the specific application of the gene sequence.[219] This provision means the abolition of absolute substance protection and the introduction of functional substance protection for sequences or partial sequences of human genes or of the genes that correspond to the structure of a natural substance or partial sequence of a human gene. Similar provisions can be found in Article 3.1d of the Italian Law Decree[220] and in the Spanish Patent Act.[221] The UK on the other hand has not expressly barred gene patents from absolute protection but had already offered purpose-bound protection for DNA sequences prior to the approval of the Biotech Directive.[222] However, in the case of European patents granted with

---

[218] See Prinz zu Waldeck und Pyrmont, 'Special Legislation for Genetic Inventions – A Violation of Article 27(1) TRIPS?' (n 170) 292.

[219] Prinz zu Waldeck und Pyrmont, 'Special Legislation for Genetic Inventions – A Violation of Article 27(1) TRIPS?' (n 170) 292. See also Franz-Josef Zimmer and Svenja Sethmann, 'Act Implementing the Directive on the Legal Protection of Biotechnological Inventions in Germany (BioPatG)' (2005) 24 Biotechnology Law Report 561. Germany's implementation of Article 9 of the Biotech Directive was delayed due to an internal debate that ended with the German Parliament's approval of more restricted protection for human DNA sequence patents in order to encourage inventors to conduct further research on DNA sequences, while keeping the same broad protection for plant and animal DNA sequences as provided in the Biotech Directive, see Bryan (n 172).

[220] Italian Law Decree (n 3) of 10 January 2006, art 3.1d; see also Prinz zu Waldeck und Pyrmont, 'Special Legislation for Genetic Inventions – A Violation of Article 27(1) TRIPS?' (n 170) 292.

[221] Spanish Act 10/2002 of 29 April 2002 incorporating the EC Directive on the protection of biotechnological inventions into the Spanish Patent Act 11/1986 of 20 March 1986.

[222] For example in *Kirin Amgen Inc v Hoechst Marion Roussel Inc* [2004] UKHL 46, [2005] RPC 9, protection was given only to the method of producing the polypeptide using the disclosed gene sequence in the manner described and not to any use of the sequence to produce erythropoietin. Similarly, protection

effect in a national country that has not adopted provisions implementing Articles 5(3) and 9 of the Biotech Directive, there is no need to disclose the function of the invention in the application and absolute substance protection should therefore be possible for human genes as well.[223]

6.4.4.1.3 Fourth question: do Articles 27 and 30 of the TRIPS Agreement affect the interpretation of Article 9 of the Biotech Directive as purpose-bound protection? The fourth question only becomes relevant if the Biotech Directive is interpreted as having an exhaustive effect on national patent law that prevents the national legislature from adopting absolute product protection for genetic inventions. This question basically asks whether Articles 27 and 30 of the TRIPS Agreement affect the interpretation of Article 9 of the Biotech Directive as purpose-bound protection.

While the TRIPS Agreement was ratified by the European Community, it has no direct effect on EU legislation.[224] The case law of the CJEU expressly precludes the possibility of testing the lawfulness of a provision of EU law against WTO agreements, even though the preparatory works of the Biotech Directive indicate that the Community legislature took the provisions of the TRIPS Agreement into account when preparing the Biotech Directive's text.[225] However, to avoid possible conflicts, the Community, when possible, applies an interpretation in keeping with the TRIPS Agreement although its provisions are not directly binding to EU courts.

After clarifying this, the CJEU noted that Articles 27 and 30 of the TRIPS Agreement related to patentability and the exceptions to the rights conferred by a patent rather than to the patent extent of protection as Article 9 of the Biotech Directive does. Notwithstanding this difference, the court explained that an interpretation of Article 9 limiting the protection it confers to situations in which the patented product performs its function does not appear to conflict unreasonably with a normal exploitation of the patent and does not 'unreasonably prejudice the

---

was not given to the DNA sequence per se in *Monsanto v Cargill* (n 193), see Cohen and others (n 103). See also Gareth Morgan, Matthew Royle and Simon Cohen, 'Cargill vs. Monsanto' (2008) 27 Biotechnology Law Report 109.

[223] See Maximilian Wilhelm Haedicke and Henrik Timmann (eds), *Patent Law: A Handbook on European and German Patent Law* (CH Beck, Hart and Nomos 2014) 671.

[224] *Monsanto v Cefetra* (n 162) points 71 and 72.

[225] See Kock (n 150), referring to *P Biret & Cie v Council* (C-94/02) [2003] ECR I-10565, paras 55 and 56 and case law cited.

legitimate interests of the patent owner, taking account of the legitimate interests of third parties', within the meaning of Article 30 of the TRIPS Agreement.[226] It thus concludes that the interpretation given of Article 9 of the Biotech Directive is not to be influenced by Articles 27 and 30 of the TRIPS Agreement.

6.4.4.2 Impact of the decision

The Monsanto decision forms part of an ongoing trend to restrict the expansion of patent rights for DNA material. In this regard, the CJEU's interpretation of Article 9 of the Biotech Directive is one of the very first manifestations of the European judiciary in this approach towards imposing heightened patentability standards for genetic inventions.[227] Nevertheless, the specific characteristics of this case, which involves dead material resulting from genetically modified soy plants, raise questions regarding the implications of the CJEU's judgment for cases where the resulting material possesses different properties. The decision of the CJEU very much takes into account the particular circumstances of the case, and thus it should be expected that future decisions will not apply the same standards if the characteristics of the case differ substantially. This interpretation would leave room for fair gene patent protection in each specific case, which should nonetheless correspond with the actual technical contribution made by the patentee.[228] In the case of gene patents, such technical contribution, according to the text of the Biotech Directive, is the invention's ability to carry out its function. Thus, although the CJEU's decision rejects absolute product protection for genes, it restricts the scope of protection no further than to the actual technical value of the invention, thus seeking to establish a balanced interpretation that serves both the interests of patentees and the general public.

For innovators within the agricultural biotech sector, purpose-bound protection would mean that patents on plant genes are no longer able to claim protection over derivative or processed products incorporating the invention unless the invention's function is also being performed. In this sense, a new form of claim drafting would need to be adopted to obtain

---

[226] *Monsanto v Cefetra* (n 162) point 76.

[227] Van Overwalle (n 189). See also Lisa A Karczewski, 'Biotechnological Gene Patent Applications: The Implications of the USPTO Written Description Requirement Guidelines on the Biotechnology Industry' (1999) 31 McGeorge Law Review 1043.

[228] Gielen (n 192).

protection on such products.[229] Furthermore, companies may also start paying more attention to alternative forms of protection such as plant breeder's rights to protection pursuant to the International Convention for the Protection of New Varieties of Plants (UPOV),[230] which provides protection for final products containing the protected plant variety. Nonetheless this decision represents an important opportunity for farmers in their battle against patents over genetically modified plants and seeds. With purpose-bound protection, farmers are no longer liable for manufacturing final products containing the patented invention unless the latter is performing its industrial application as indicated by the patent applicant.

For patentees in general, the decision removes a broad layer of protection that may erode the value of key patents within a company's portfolio.[231] Article 9 of the Biotech Directive breaks with the principle of absolute product protection prevailing in other technical fields and approximates gene patent protection to the protection conferred by use or method patents.[232] It has been claimed that this ruling undermines the efforts for global intellectual harmonization by creating a lucrative loophole for patent infringement because growers could potentially circumvent gene patents over genetically modified crops by growing the crops in a country where the gene is not patented, and then importing the product into a country where the patent is in force with impunity.[233] Nevertheless, as long as the imported product does not infringe on the patent under Article 9 of the Biotech Directive, the fact that an invention is not protected in a certain country cannot be interpreted as an opportunity to easily infringe on the patentee's rights.

On the other hand, the Biotech Directive's conception of genetic products as inventions prevents those patents from being conceived as conferring purpose-limited protection of the type available for second and subsequent medical uses of known products containing claims limited to

---

229 See Morgan and Haile (n 203).

230 See Vid Mohan-Ram, Richard Peet and Philippe Vlaemminck, 'Biotech Patent Infringement in Europe: The "Functionality" Gatekeeper' (2011) 10 J The John Marshall Review of Intellectual Property 1540.

231 See Morgan and Haile (n 203).

232 See Prinz zu Waldeck und Pyrmont, 'Special Legislation for Genetic Inventions – A Violation of Article 27(1) TRIPS?' (n 170) 293; Craig C Carpenter, 'Seeds of Doubt: The European Court of Justice's Decision in *Monsanto v Cefetra* and the Effect on European Biotechnology Patent Law' [2010] The International Lawyer 1189.

233 Carpenter (n 232).

one or more specific indications of a drug, so that third parties are lawfully entitled to use or make (etc.) any non-patented indications of the drug. The CJEU's interpretation of Article 9 of the Biotech Directive potentially supports a more restrictive definition of patent scope via a different test of patent infringement, according to which a patent for a genetic product will be infringed if and only if a defendant makes or uses the product in a form in which it remains susceptible to industrial application in the manner described by the patentee in the specification. In contrast, the test of infringement that has been applied with respect to purpose-limited products[234] is whether the defendant knew or could reasonably foresee that the product would be intentionally used for the patented indication, a test that potentially narrows third parties' scope for manoeuvre.

Furthermore, the Monsanto decision may also have an impact on the viability of patents claiming isolated DNA or RNA sequences used as reagents, including reagents used in diagnostic methods such as gene tests and DNA chips.[235] These nucleotides do not perform their function in a reagent vial or kit, thus patent claims directed to this subject matter may not be enforceable in view of the CJEU's judgment.[236] On the other hand, the need to prove that the infringing product is performing the patent's intended utility will increase litigation costs on the plaintiff's side, which may discourage patentees such as medium-size companies from asserting their patent rights. Furthermore, this fact would also help to deter firms from starting litigation proceedings for strategic purposes only. In this regard, although discouraging patent owners from enforcing their exclusive rights is potentially prejudicial for patentees in this area, it represents a significant step towards public access to genetic diagnostic and test tools.

Finally, with regard to the role of Article 5(3) of the Biotech Directive in interpreting purpose-bound protection, the CJEU clarified that the function referred to in Article 9 is the industrial applicability disclosed by

---

[234] In the UK, see *Warner-Lambert Company, LLC v Actavis Group PTC EHF & Ors* [2015] EWCA Civ 556. In Germany, see *Cistus Incanus* (OLG Düsseldorf) [2013] Case I-2 U 54/11; *Chronic Hepatitis C Treatment* (Landgericht Düsseldorf) [2013] Case 4A O 145/12. Both German decisions refer to the notion of 'manifest making-up' or 'obvious arrangement' ('sinfällige Herrichtung') that may result from the particular configuration of the substance or article, or in the addition of a package's leaflet.

[235] See Mohan-Ram, Peet and Vlaemminck (n 230).

[236] Ibid.

the applicant and not the biological function.[237] Thus, the interplay between both articles and related recitals gives the requirement of industrial application a fundamental role in the patentability of gene patents. Furthermore, although Article 5 of the Biotech Directive only seems to concern the human body and its parts, the Monsanto decision subtly extends the scope of the requirement in Article 5(3) concerning human genes and sequences thereof to cover genetic inventions in general. The CJEU decision reaffirms that the requirement of Article 9 of the Biotech Directive applies to human, animal and plant genetic sequences and therefore, by referring to the industrial application disclosed according to Article 5(3), this decision implies that the statement on function in the patent application required in Article 5(3) would also apply to non-human genes and gene sequences,[238] thus extending the impact of the disclosure of industrial application requirement to all genetic inventions.

6.5 CONCLUDING REMARKS

As discussed in the second section of this chapter, a clear delimitation of the scope of the patent is determinant for effectively defining the extent of the rights of the patentee and the limitations that such rights impose to competitors. However, the special characteristics of fast-developing industries like genetics, where many inventions claim pioneer status, pose problems to the application of general patent provisions on patent scope. In particular, inventors' races for obtaining priority and the broadest possible scope of patent protection gave rise to patent claims that did not correspond with the invention's technical contribution to the art. In order to ensure that the scope of genetic patents is limited to their actual practical value, Article 9 of the Biotech Directive restricted the scope of protection of all patents concerning genetic material to their ability to perform their intended function. This approach prevents the grant of absolute product protection, at least within the context of human DNA sequences as such, and thus is compatible with those economic theories on the size of patent scope that support narrowing the scope of protection granted to inventions in groundbreaking scientific fields.

Article 9 of the Biotech Directive simply introduces specific requirements for genetic inventions and thus is applied in addition to the general European patent law rules on the determination of patent scope. In fact,

[237] Krauss and Takenaka (n 33).
[238] Haedicke and Timmann (n 223) 675.

the interpretation of Article 9 as purpose-bound protection has been discussed by the CJEU in *Monsanto v Cefetra.* In line with the Parliament's view, the CJEU has made clear with this decision that Article 9 is to be interpreted as limiting the scope of protection of gene patents to their ability to perform their desired function. Furthermore, the CJEU held that the function referred to in Article 9 is the industrial applicability disclosed by the applicant according to Article 5(3) of the Biotech Directive, which concerns patents over human genes and fragments thereof. Therefore, although the suitability of applying such an approach to non-human gene patents is questionable, by referring to Article 5(3) of the Biotech Directive, the Monsanto decision implicitly extends the scope of the requirement in Article 5(3) concerning human genes to cover genetic inventions in general. Thus, together with Article 5(3) and related recitals, Article 9 forms part of an ongoing trend towards restricting the patentability of genetic material to the extent that they are usable in practice.

7. Human gene patents, patent clusters and innovative progress

7.1 INTRODUCTION

In addition to the continuous increase in the number of patents concerning biotechnological inventions since the early 1990s, today more and more products incorporate not just a single new invention but a combination of many different components, each of which may be the subject of one or more patents. In industries like biotechnology where research is cumulative, this situation poses difficulties to the operation of the patent system. In particular, there is the concern that extensive patenting and over fragmentation of rights may give rise to clusters of overlapping patents that could pose barriers to the development of subsequent innovations. It has been argued that if such patent floods emerge, a restrictive exercise of patentees' exclusive rights would impede access to essential technologies, discourage further research and restrain future innovation. In this regard, different mechanisms to address the consequences of this situation have been proposed, and in some cases successfully implemented. However, none of the existing alternatives offers an overarching solution to restrict the potentially negative impact of extensive patenting on subsequent innovation in gene technologies.

This chapter argues that notwithstanding the enormous significance of ex post models in improving access to important genetic innovations, the Biotech Directive's strict interpretation of the requirement of industrial application for genetic inventions, in the sense that it poses a higher barrier to patentability and limits the scope of protection granted to these types of inventions, can assist patent authorities in avoiding extensive patenting of gene sequences that do not provide a useful contribution to the art that a licensee can actually exploit, but only add extra layers to the existing mass of genetic patents.

Within that context, the chapter first analyses the background and effects of extensive patenting in the progress of innovation in gene technologies. Second, it explores the related 'theory of the anticommons', its main arguments, potential implications and practical limitations, and then discusses the possibility of an anticommons situation in the field of gene-based research tools. Finally, the chapter analyses some possible solutions to the problems associated with patent thicketing, their characteristics and limitations, and how the strict European approach towards the implementation of the industrial application requirement (that is applied prior to other promising ex post access mechanism) can help to impede the formation of unduly restrictive patent conglomerates by limiting the number of broad patents lacking practical utility.

7.2 PATENT CLUSTERS IN BIOTECHNOLOGY

7.2.1 Strengthening Patent Systems in Europe

Over the last three decades governments worldwide have adopted policies aimed at strengthening patent systems. In particular, reforms have been mainly focused on two objectives, encouraging innovators to use the patent system and reinforcing the rights of patent holders.[1] Following the US example, Europe has gradually developed innovation policies that reflect a notably strong pro-patenting approach; especially in emerging industries like biotechnology. Relying on the assumption that patents are essential for boosting innovation, European governments have adopted measures to promote the utilization of the patent system. With regard to biotechnology, governments have implemented policies directed at:

- making patents available for new subject matter, such as genetically engineered life forms or isolated human genes, and new types of applicants like universities and other publicly funded research institutions
- extending the term of patent protection and
- establishing lower patenting standards.

7.2.1.1 New patentable subject matter

Partly in response to the perceived demands from emerging fields of research and technology such as human genetics, since the 1980s there

---

[1] Policies aimed at strengthening patent systems for the purpose of spurring innovation find support under the traditional conception of patents as public trade-offs in which patents incentivize research and knowledge disclosure, but at the social costs of restricting access to the invention during the life of the patent, see Nancy T Gallini, 'The Economics of Patents: Lessons from Recent U.S. Patent Reform' (2002) 16 The Journal of Economic Perspectives 131.

has been a tendency towards broadening the definition of patentable subject matter. In this regard, European countries have developed policies and rules to consolidate the inclusion of living material as patentable subject matter.[2] In consequence, the EPO and national patent offices started to frequently issue patents on inventions concerning newly discovered genetic material such as isolated human gene fragments (see Chapter 5).

Nonetheless, policies extending patentability in these areas have received strong criticism, especially in the case of gene-based technologies.[3] As discussed in Chapter 2, patents are widely seen as an effective mechanism for promoting innovation in the particular field of biotechnology. However, the question of whether biotech inventions should be susceptible to patent protection is still the focus of debate. In particular, disputes continue to arise regarding the types of biotech inventions that merit patent protection.[4] In this sense, while patents concerning the use of enzymes in the paper industry hardly attract much attention, patents over isolated human gene fragments for the development of diagnostic kits are often the focus of controversy.

7.2.1.2 Universities and public research institutions as patent owners

Until the 1980s the biotechnology field resembled a commons model in which governments encouraged the free flow of information about new scientific advances such as genetic discoveries and the results from publicly funded research used to be immediately published.[5] Public institutions owned few patents and the technologies covered by those patents were not commercialized.[6] An example would be the discovery of

---

[2] For instance, the Directive 98/44/EC of the European Parliament and of the Council of 6 July 1998 on the legal protection of biotechnological inventions [1998] OJ L213/13; Guidelines for Examination in the European Patent Office (2002) Part G.II.2, 3.1.

[3] See Gallini (n 1).

[4] See OECD, 'Genetic Inventions, Intellectual Property Rights and Licensing Practices, Report of a Workshop Organized by the OECD Working Party on Biotechnology' (OECD 2002) 11.

[5] Daniel Zatorski, 'The Tragedy of the Anticommons in Biotechnology' (DPhil thesis, The Jagiellonian University in Krakow Intellectual Property Law Institute 2011) 30.

[6] Rebecca S Eisenberg, 'Patents and the Progress of Science: Exclusive Rights and Experimental Use' (1989) 56 U Chi L Rev 1017; Michael S Mireles, 'An Examination of Patents, Licensing, Research Tools, and the Tragedy of the Anticommons in Biotechnology Innovation' (2004) 38 University of Michigan Journal of Law Reform 141.

the monoclonal antibody by Cesar Milstein and George Kohler in 1975, which was not patented and was freely accessible by the public.[7] However this is no longer the case. Claims about the existence of a supposed 'European paradox' in academic research – according to which many European countries could hold prominent worldwide positions in terms of scientific achievements, but would not be able to transform them into technological advantages – have triggered policy measures aimed at facilitating the transformation of academic scientific research into marketable technical products.[8] In this regard, patents have started to be considered as a good means for exploiting the commercial and social benefits of discoveries resulting from publicly funded academic research; and thus, for bringing scientific knowledge into practice.

Traditionally, the prestige and value of academic researchers were mainly based upon peers' recognition of their contributions in advancing scientific knowledge.[9] Publicly supported research was presumptively placed in the public domain, while privately funded research was typically appropriated as intellectual property. However, during the past decades, the remarkable value of upstream research projects carried out by universities (especially in biomedical sciences) has helped to attract private investment.[10] As a result, commercial biotechnology firms have emerged between the upstream research of academic laboratories and the targeted downstream product development of private firms, which has made it difficult to maintain the previous boundaries between public and private research in biotech sciences.[11]

Currently upstream research in biotechnology is increasingly likely to be private in one or more senses of the term, for instance:

---

[7] Michael Heller, *The Gridlock Economy: How Too Much Ownership Wrecks Markets, Stops Innovation, and Costs Lives* (Basic Books 2010) 58.

[8] C Franzoni and F Lissoni, 'Academic entrepreneurship, patents, and spin-offs: critical issues and lessons for Europe' [2006] Centro di Ricerca sui Processi di Innovazione e Internazionalizzazione (CESPRI); Francesco Lissoni and others, 'Academic Patenting in Europe: New Evidence from the KEINS Database' (2008) 16 Research Evaluation 87.

[9] Recognition is usually presented in the form of references or citations to prior authors' claims regarding discoveries or ideas. Being credited by other researchers is very important for a scientist's career.

[10] Rebecca S Eisenberg, 'Intellectual Property Issues in Genomics' (1996) 14 Trends in Biotechnology 302; Paula E Stephan and Stephen S Everhart, 'The Changing Rewards to Science: The Case of Biotechnology' (1998) 10 Small Business Economics 141.

[11] Michael A Heller and Rebecca S Eisenberg, 'Can Patents Deter Innovation? The Anticommons in Biomedical Research' (1998) 280 Science 698.

- supported by private funds
- carried out in a private institution or
- privately appropriated through patents, trade secrecy or agreements that restrict the use of materials and data.[12]

Researchers in the public and private sectors are often working on the same problems, whether competitively or collaboratively,[13] and the incentives that govern private property rights have been progressively incorporated into academic and other types of public research institutions. Today the value of an investigation often depends in great part on the degree to which a researcher can generate an invention.[14] The prevailing wisdom is that institutions performing publicly sponsored research should patent their discoveries to promote commercial development,[15] which makes universities increasingly eager to obtain patent protection for their research results. In consequence, researchers now increasingly seek to be co-inventors of patents and receive royalties from licensing their inventions.

University patents came to the attention of the European academic community, policy makers and the general public as a consequence of the impressive growth in the number of patent applications by US universities after the introduction of the Bayh–Dole Act in 1980.[16] However, the history and organization of the US system differs in many ways from the European one.[17] In several European jurisdictions the existence of the

---

[12] Ibid.

[13] Eisenberg, 'Intellectual Property Issues in Genomics' (n 10).

[14] Fiona Murray and Scott Stern, 'When Ideas Are Not Free: The Impact of Patents on Scientific Research' (2006) 7 Innovation Policy and the Economy 33.

[15] Eisenberg, 'Intellectual Property Issues in Genomics' (n 10).

[16] Bayh–Dole Act 35 USC 200–212. For further reading in US academic patenting see Rebecca Henderson, Adam B Jaffe and Manuel Trajtenberg, 'Universities as a Source of Commercial Technology: A Detailed Analysis of University Patenting, 1965–1988' (1998) 80 Review of Economics and Statistics 119; D Mowery and others, 'The Growth of Patenting and Licensing by U.S. Universities: An Assessment of the Effects of the Bayh–Dole Act of 1980' (2001) 30 Research Policy 99; D Mowery and A Ziedonis, 'Academic Patent Quality and Quantity Before and After the Bayh–Dole Act in the United States' (2002) 31 Research Policy 399. For the UK see UKIPO Lambert Working Group on IP, Lambert Review of Business–University Collaboration (Final report, UKIPO 2003) and subsequent Lambert Toolkit (UKIPO 2005).

[17] The US university system has been, since its early boom in the first half of the nineteenth century, a heterogeneous collection of autonomous institutions. Their faculty members were not subject to their students' control, nor were they civil servants paid by the state, as in most European countries. Attempts to

so-called 'professor's privilege', which exempts academic personnel from attributing the rights over the inventions they generate to their employers, gives academic individuals the power to decide whether to patent, commercialize, or license their own inventions. Based on the argument that the professor's privilege discourages academic scientists to commercialize their intellectual property rights, this exemption was abolished in some European countries including Austria, Denmark, Finland, Germany and Norway.[18] In other countries, like France, Ireland and the UK, governments have adopted institutional ownership rights over academic inventions and (or) guidelines for incentivizing university patenting and fostering more consistent practices in IP management.[19] Jointly, these reforms have led to a notable increase in university patenting in most European countries.

7.2.1.3 Extended term of patent protection

With the introduction of the TRIPS Agreement in 1995 the term of patent protection was extended to a minimum of 20 years in all WTO Member States.[20] Thus, all European countries that were members of the WTO were required to adopt the new term of patent protection if they had not done so yet. Furthermore, since 1993 supplementary protection certificates (SPCs) are available for medicinal products, which allowed this period to be extended for a maximum of five years from the end of the lawful term of the basic patent for a period equal to the period which elapsed between the date on which the application for a basic patent was lodged and the date of the first authorization to place the product on the

---

centralize the university system have always failed, even at times of financial difficulties. Nowadays, autonomy is one of the greatest strengths of the US universities, and this is also the main background reason for their transformation into entrepreneurial organizations, see Franzoni and Lissoni (n 8). See also Henderson, Jaffe and Trajtenberg (n 16).

[18] Francesco Lissoni and others, 'Academic Patenting and the Professor's Privilege: Evidence on Denmark from the KEINS Database' (2009) 36 Science and Public Policy 595.

[19] Aldo Geuna and Federica Rossi, 'Changes to University IPR Regulations in Europe and the Impact on Academic Patenting' (2011) 40 Special Issue: 30 Years After Bayh–Dole: Reassessing Academic Entrepreneurship 1068; UKIPO, 'Intellectual Asset Management for Universities' (UKIPO 2014).

[20] Agreement on Trade-Related Aspects of Intellectual Property Rights of 15 April 1994, art 33.

market in the Community.[21] SPCs were established to guarantee sufficient protection for the development of medicinal products in the EU and to help compensate owners for the lengthy time period necessary to obtain regulatory marketing approval. In the case of biotechnology, diagnostic tests, biopharmaceuticals and other biomedical products are entitled to apply for such supplementary protection.

7.2.1.4 Lower standards for patentability

In new industries like biotechnology where there are a large number of inventions, the distinction between what constitutes an invention and what is a mere enhancement of an acknowledged technology is becoming increasingly difficult to discern.[22] Thus, the evaluation of patent law requirements is often made in accordance with flexible criteria that sometimes lead to the granting of patents over pure discoveries or minor or trivial developments.[23] Furthermore, applicants are in many cases successful in their attempts to broaden patent claims as much as possible in order to secure stronger scopes of protection that do not always correspond with the actual utility of the invention (see section 6.2).

In order to address some of the problems created by low patentability standards, the EPO adopted a 'Raising the Bar' policy directed to ensure that 'the EPO grant patents only for innovations having sufficient inventive merit and meeting the needs of society'.[24] This same issue has also become a common concern among national patent authorities. However, not all the measures proposed have been successfully implemented and concerns still remain.

7.2.2 The Result: Patent Clusters, Royalty Stacking, Patent Holdup and Defensive Patenting

The patent law policies and reforms discussed above have resulted in a notable growth in the total number of biological patents and patents covering a single technology. Moreover, the new pro-patenting scenario

---

[21] Regulation (EC) No 469/2009 of the European Parliament and of the Council of 6 May 2009 concerning the supplementary protection certificate for medicinal products (Codified version) OJ L152/1, art 13.

[22] See UKIPO, 'Patent thickets: an Overview' (UKIPO 2011) 3.

[23] Carlos M Correa, 'Efforts to Raise the Bar in Patent Examination Need to be Supported' (2012) 43 IIC 747.

[24] EPO, 'Quality over quantity: on course to raise the bar' (2008) <http://www.epo.org/about-us/office/annual-report/2008/focus.html> accessed 10 October 2012.

has also helped to encourage strategic patenting and increase the amount of upstream patents owned by universities and other public research institutions.

In the last few years, the numbers of European patent filings and granted patents have risen in the biotech sector in general. Part of this growth can be explained by increased investment in research and development. However, a large part of this increase in patent activity is due to existing policies and strategic patenting behaviours aimed at reinforcing the bargaining position of patent holders and avoiding competition.[25] Between 2002 and 2011 there has been a steady rise in the number of European biotech patents.[26] In 2011 the EPO received around 5900 applications compared to the 5700 that were filed in 2002. However, the number of granted patents has shown a more pronounced increase. In 2011 approximately 2100 patents were granted while in 2002 the EPO decided to grant patents over approximately 1000 inventions.

The number of so-called 'submarine patents', which are patents the issuance of which is intentionally delayed by the applicant, and 'reach-through patent claims', which give the owners of patents over upstream technologies rights in subsequent downstream discoveries, has also grown.[27] Moreover, along with the general rise in the number of patents, there have also been numerous cases where an individual patent does not cover the entire set of inventions that form a specific product, but represents only a piece in a puzzle of patents covering a technical product or process.[28] In contrast with the one-to-one correspondence between products and patents that might be expected, today there are many cases where hundreds of patents, which frequently overlap with each other, cover the same product. Moreover, those conglomerates of patents over a

---

[25] Markus Reitzig, 'The Private Values of Thicket Patenting: An Empirical Study Across Industries and Patent Types' (DRUID Summer Conference on 'Industrial Dynamics of the New and Old Economy – who is embracing whom?', Copenhagen, June 2002).

[26] EPO Annual Report 2011, Statistics.

[27] For example, research tool patents over, for example, markers, assays, receptors or transgenic animals, increasingly claim to cover products identified by the patented tool or method. If such a claim is granted, patent owners can demand royalties on the sale of a product found with the help of their research tool. Since many different patented research tools must be used in the development of drugs, reach-through claims increase royalty stacking, see OECD, 'Genetic Inventions, Intellectual Property Rights and Licensing Practices, Report of a Workshop Organized by the OECD Working Party on Biotechnology' (n 4) 15–16.

[28] Reitzig (n 25).

single technology rarely belong to a single owner, but ownership over those patent rights is usually divided among multiple owners.

European universities are also increasingly active in terms of IP commercialization. Today intellectual property rights over upstream inventions resulting from publicly funded research belong to universities in many countries including Austria, Belgium, France, Germany, Netherlands, Norway, Portugal, Spain and the UK.[29] However, even though the number of patents owned by European universities and other public institutions has increased in the last decade,[30] the level of academic patenting in Europe still lags behind its US counterpart.[31] Nevertheless, the situation has raised concerns among academics and other stakeholders. In particular, fears have been expressed about the impact of university patenting in future research.

In discrete technologies patent holders can independently produce their goods and are not forced to enter collaborations with other technology holders whereas in cumulative technologies the opposite is true.[32] In fields like biotechnology, where research is cumulative and builds upon previous discoveries, the increasing number of overlapping patents has contributed to the development of so-called 'patent clusters'. The fact that those patents do not always fully comply with the traditional patent law requirements has also facilitated the formation of such conglomerates, creating difficulties for innovators. In this regard, the increase of

[29] OECD, Compendium of Patent Statistics (2008) 25.

[30] Magnus Klofsten and Dylan Jones-Evans, 'Comparing Academic Entrepreneurship in Europe – The Case of Sweden and Ireland' (2000) 14 Small Business Economics 299; OECD, Compendium of Patent Statistics (n 29) 4 and 25.

[31] European academic patents differ from US university patents to the extent that, while both protect inventions developed by academic scientists, the former tend to be owned by companies and, to a lesser extent, public research organizations. In contrast, US university patents are owned by the universities themselves. However, with the encouragement of governments, European universities have become more aggressive in trying to retain the intellectual property rights over their scientists' inventions. While in the past they left their scientists to dispose of such rights freely or allowed private sponsors to retain all of the intellectual property, nowadays these universities tend to impose disclosure norms. As a result, they have been better equipped to build up patent portfolios for commercial exploitation, see Robin Cowin and others, 'Policy Options for the Improvement of the European Patent System' (2007) Scientific Technology Options Assessment (STOA) of the European Parliament 20.

[32] Reitzig (n 25).

patents with broad claims and patents in the hands of public institutions has exacerbated the problem.

Patent clusters or floods are also usually referred to as 'patent thickets', a term that originated in litigation in the 1970s regarding Xerox's dominance of a portion of the photocopier industry.[33] In 2000, economist Carl Shapiro re-introduced the term in academic discourse by defining patent thickets as 'a dense web of overlapping intellectual property rights that a company must navigate its way through in order to actually commercialize new technology'.[34] In essence, patent thicket is a descriptive term that refers to the problems that individuals may face when attempting to enter (or innovate in) a technology market that is flooded with upstream patent rights. In this scenario, patents cause regulatory blockage in the form of conglomerates of pre-existing patents (and pending patents), which impose constraints on new innovators wishing to enter the market. Due to the negative connotation that this term carries, this book refers to the phenomenon of extensive patenting and fragmentation of rights as patent clusters, since such scenarios do not always result in insurmountable obstacles for the progress of innovation.

There are four main types of relationships between patents, namely:

(1) blocking
(2) complementary
(3) independent and
(4) substitute.

The evaluation of the positive and negative effects of a certain combination of patents would be much easier if all of the potential relationships between patents fell entirely into one these four categories. However, this is rarely the case. When two patents (A and B) block each other, the owner of patent B cannot practise the invention without a licence from the owner of patent A.[35] On the other hand, two patents that provide an

---

[33] *Re Xerox Corp*, 86 FTC 364 (1975); *SCM Corp v Xerox Corp*, 645 F2d 1195 (2d Cir 1981).

[34] Carl Shapiro, 'Navigating the Patent Thicket: Cross Licenses, Patent Pools, and Standard-Setting' in Adam B Jaffe, Josh Lerner and Scott Stern (eds), *Innovation Policy and the Economy* (MIT Press 2001) 119.

[35] See Roger B Andewelt, 'Analysis of Patent Pools under the Antitrust Laws' (1984) 53 Antitrust Law Journal 611; Robert P Merges and Richard R Nelson, 'On The Complex Economics of Patent Scope' (1990) 90 Columbia Law Review 839.

additional benefit when used in combination are said to be complementary. Unlike blocking patents, complementary patents can each be practised independently without requiring a licence for the other patent. In contrast, independent patents are those that do different things and do not need to use other patents to perform their functions. Finally, two patents that perform substantially identical functions or fulfill the same role but can be practised independently are considered to be substitutes.[36] While substitute and independent patents do not usually impose barriers to new innovators, blocking and complementary patents are likely to do so.

According to Shapiro's metaphor, the process of research and development is comparable to the continuous extension of a pyramid through the addition of new building blocks at the top.[37] In the case of biotechnology, firms increasingly protect their contributions to this pyramid with overlapping patents.[38] Whether these patents are complementary or block each other, newcomers to the pyramid need to obtain licences for many or all the patents covering a particular technology that are essential to their research, which entails negotiating with all the patent holders concerned.

The principal problems associated with patent clusters can be classified into four main issues:

(1) royalty stacking
(2) patent holdup
(3) strategic patenting and
(4) the related anticommons problem in sequential innovation.

In practice, these problems often occur simultaneously or as a consequence of another's effect.

7.2.2.1 Royalty stacking

Royalty stacking arises when the combination of royalty fees that the developer of a new product needs to pay in order to cut through a patent cluster finally makes the product itself unprofitable. Despite the new wording, the economic logic behind royalty stacking is based on the 'complements problem' first studied by Augustin Cournot in 1838. Cournot showed that consumers benefit when all products that are complementary from a demand viewpoint are produced and marketed by

---

[36] Ibid.
[37] Shapiro (n 34) 119.
[38] Heller and Eisenberg (n 11).

a single firm.[39] However, in the case of patent rights, the complements problem is far more complex.

Obtaining licences from multiple patent owners may entail long and expensive negotiations, high economic risks and sometimes litigation costs. Moreover, since all the patents in a research area rarely belong to a single owner, businesses may find it difficult or even impossible to know with whom they are in conflict, or whom they should approach for a licence. While the precise extent of these problems remains unclear, empirical evidence has demonstrated that royalty stacking is far more than a theoretical possibility.[40] In the field of genetics, current trends towards disintegration in the ownership of patent rights have revived discussions about the complements problem in innovative research and product development.

For example, although a database of gene patents is a useful resource for improving transparency, granting too many property rights around isolated gene fragments could result in costly transactions to bundle licences together before a firm can have an effective right to develop new gene-based technologies.[41] When companies engage in a new project, an initial study of the patent landscape can sometimes reveal that there are dozens, sometimes more than 100 patents to consider.[42] This was the case of the Malaria Vaccine Initiative or the Golden Rice case where a freedom-to-operate survey initially uncovered 70 patents belonging to 32 different companies and universities.[43] In the long run, the sum of payable royalty fees may stop innovators from developing certain products, or divert resources to less promising projects,[44] if that cost sufficiently reduces the profits from designing and manufacturing new technologies.

---

[39] Damien Geradin, Anne Layne-Farrar and A Jorge Padilla, 'The Complements Problem within Standard Setting: Assessing the Evidence on Royalty Stacking' (2008) 14 BUJ Sci & Tech L.

[40] Mark A Lemley and Carl Shapiro, 'Patent Holdup and Royalty Stacking' (2007) 85 Texas Law Review 1991.

[41] Heller and Eisenberg (n 11).

[42] See OECD, 'Genetic Inventions, Intellectual Property Rights and Licensing Practices, Report of a Workshop Organized by the OECD Working Party on Biotechnology' (n 4) 61; Birgit Verbeure and others, 'Patent Pools and Diagnostic Testing' (2006) 24 TRENDS in Biotechnology 115.

[43] Ibid.

[44] Heller and Eisenberg (n 11).

7.2.2.2 Patent holdup

Holdup problems are common to all industries where patents exist. However, this problem is more frequent and complex in sectors that host heavy patent conglomerates. Holdup problems may appear, for instance, in cases where a drug manufacturer that has designed and started to produce a given product suddenly realizes that he may be infringing on one or more patents. When this happens, the negotiating positions of the manufacturer and the patentees are unlikely to be balanced. Moreover, in the field of genetics the lack of substitutes for certain biomedical discoveries like patented gene sequences or receptors may increase the imbalance between patent holders.[45] If the manufacturer decides not to obtain those licenses, his product may be found to be infringing on others' rights, even if the manufacturer was not aware of the existence of these patents. Thus, in order to avoid risk the manufacturer will probably opt to acquire the necessary licences, and in this case the patentee could and probably would seek greater royalties and impose stricter conditions since a refusal to grant a licence could be highly damaging to the new manufacturer.

7.2.2.3 Defensive patenting

In contrast, strategic or defensive patenting refers to situations in which firms file patent applications only in order to strengthen their bargaining positions and extract benefits from competitors. In this regard, even though strong patents may facilitate the transfer of technology, they may also facilitate anticompetitive behaviour. Survey evidence suggests that firms in many industries have not increased their reliance on patents for appropriating the returns from research and development investments over the decade of the 1980s (see section 2.2.1). However, the number of patents has continued to rise.

Today many companies aggressively seek to build large patent portfolios and use them as 'bargaining chips' to negotiate with competitors.[46] For many firms, the only practical response to unintentional and sometimes unavoidable patent infringement is to file hundreds of patents each year in order to have something to trade with during cross-licensing negotiations.[47] Clear examples of strategic patenting are the so-called

---

[45] Ibid.

[46] Bronwyn H Hall and Rosemarie Ham Ziedonis, 'The Patent Paradox Revisited: An Empirical Study of Patenting in the U.S. Semiconductor Industry, 1979–1995' (2001) 32 The RAND Journal of Economics 101.

[47] Gavin Clarkson, 'Objective Identification of Patent Thickets: A Network Analytic Approach' (DPhil thesis, Cambridge, MA: Harvard University 2004) 7.

'patent trolls', which are organizations that amass patent rights for the purpose of conducting opportunistic patent litigation against alleged infringers. As Turner describes them, 'trolls do not reduce technology to practice yet close it off to use by complementary technologies'.[48] Thus, the existence of clusters of overlapping patents encourage not only defensive patenting but also the formation of patent trolls seeking to obtain benefits from those firms that perform research cumulatively.

7.3 THE ANTICOMMONS OF BIOTECHNOLOGY RESEARCH

7.3.1 Background

As explained earlier in this chapter, in industries like biotechnology where research is cumulative and innovations build upon existing knowledge, patent clusters can act as barriers that hinder innovation rather than encourage it. However, although the existence of such patent conglomerates might impede an optimal functioning of the patent system, to what extent they are able to stifle innovation and research is not yet entirely ascertained. Within the context of patent property rights, the so-called 'theory of the anticommons' refers to the idea that a massive increase in the number of patents (or patent floods) will hamper the development of new inventions and thus innovation. In contrast, the counterarguments to the anticommons proposition are that the theory contains important shortcomings in addition to the relative lack of empirical evidence. In order to assess whether patent clusters in the biotechnology industry can create a situation of tragedy in the levels of innovation and research, it is important to consider the reasoning, implications and limitations of the anticommons theory.

In 1998, by reversing the definition of a commons, Heller described what he called an 'anticommons situation'. In an anticommons multiple owners are each endowed with the right to exclude others from a scarce resource while nobody has an effective privilege of use.[49] By contrast, in a 'commons situation' multiple owners have the right to use a given

---

[48] John L Turner, 'Patent Thickets, Trolls and Unproductive Entrepreneurship' (2011) <http://ssrn.com/abstract=1916798> accessed 19 September 2012.

[49] Michael A Heller, 'The Tragedy of the Anticommons: Property in the Transition from Marx to Markets' (1998) 111 Harvard Law Review 621.

resource and no one has the right to exclude others.[50] Thus, although in a commons nobody has the right or duty to exclude others from using the shared resource (open access), neither can anyone sell or administer such resource.

Subsequently, the idea of a 'tragedy of the anticommons' was developed by mirroring the so-called 'tragedy of the commons', which was first proposed by Hardin in the 1960s. The tragedy of the commons is based around the 'rational choice theory', which supports the notion that man seeks to maximize utility in all areas of life. From a socio-economic perspective, the rational choice theory explains that people maximize the utility of a given good because maximization is rational.[51] Building upon this assumption, Hardin describes how freedom in the commons would result in the destruction of a given resource, that is, how individuals with access to the commons would fail to consider the social cost of their actions. The example given by Hardin to illustrate his theory is an open access pasture, in which each rational utility maximizer herdsman would try to keep as many cattle as possible on the commons since each of the individual herdsmen with privileges to use the pasture would obtain more advantages than disadvantages by adding cows to graze on the common resource.[52]

On the other side of the spectrum, Heller proposes a different tragedy. By using the example of empty Moscow storefronts, Heller illustrates how multiple rights of exclusion over a scarce resource could lead to a tragedy of underuse.[53] In a commons situation misalignments may appear due to externalities not captured in the calculus of interests of the

---

[50] Ibid.

[51] John Scott, 'Rational Choice Theory' in Gary Browning, Abigail Halcli and Frank Webster (eds), *Understanding Contemporary Society: Theories of the Present* (SAGE 2000) 126.

[52] Garrett Hardin, 'The Tragedy of the Commons' (1968) 162 Science 1243. The fundamental ideas of this theory have been used to explain other environmental problems like pollution, overexploitation of fisheries or destruction of national parks, but also to justify the transition from public to private property rights, see for example Harold Demsetz, 'Toward a Theory of Property Rights' (1967) 57 American Economic Review 347.

[53] Heller, 'The Tragedy of the Anticommons: Property in the Transition from Marx to Markets' (n 49). Also relating to this idea is Locke's spoliation proviso, which prohibits spoilage due to excess appropriation, and within the context of intellectual property may provide an answer to those challenging cases where patent rights promote wasteful over-appropriation and thus do not leave 'enough' for others, see Robert P Merges, *Justifying Intellectual Property* (Harvard University Press 2011) 56–59.

individual users.[54] However, the tragedy of the anticommons may occur because multiple holders of exclusion rights do not fully internalize the costs created by the enforcement of their rights to exclude others. In other words, Heller's theory suggests that although over exploitation of available resources is far from complying with the conditions of optimal use, neither is a situation of under exploitation. Therefore, like two sides of the same coin, both commons and anticommons systems of property, if taken to the limit, could provoke a sub-optimal utilization of a given resource.

7.3.2 The Tragedy of the Anticommons in Gene Patenting: Implications and Shortcomings

Although the anticommons theory had been to some extent discussed in the property literature, Heller's article has had a major impact on intellectual property discussions, especially in relation to the problem of patent clusters in software and gene technologies. Since a patent grants its owner the right to exclude others from using the patented invention, the theory of the anticommons helps to explain some of the problems encountered by the patent system due to the creation of clusters of overlapping patents. In essence, this theory challenges the premise that stronger patent protection will always foster innovation. It builds on the idea that excessive patenting can inhibit the free flow and diffusion of scientific knowledge and the ability of researchers to build cumulatively on each other's discoveries; and thus the expansion of patent rights privatizing the scientific commons would ultimately limit scientific progress. For example, it is argued that in technological fields where dense patent floods have emerged and patentees refuse to license their rights, royalty stacking problems may discourage the development of promising and innovative ideas.

Over fragmentation of patent rights usually generates additional costs to the practical functioning of the patent system. This is because the reunification of fragmented rights generally involves transaction and strategic costs of a great magnitude, including the transaction costs of rearranging entitlements, heterogeneous interests of owners and cognitive biases among researchers that lead to over estimating their intellectual

---

[54] Francesco Parisi, Norbert Schulz and Ben Depoorter, 'Duality in Property: Commons and Anticommons' (2005) 25 International Review of Law and Economics, 578; Ben Depoorter and Sven Vanneste, 'Putting Humpty Dumpty Back Together: Experimental Evidence of Anticommons Tragedies' (2006) 3 Journal of Law, Economics & Policy 1.

property assets.[55] In this sense, the tragedy of the anticommons describes a hypothetical situation in which the negative externalities created by multiple rights of exclusion over blocking or complementary technologies provoke the waste of resources that may otherwise help to promote innovation and development.

The anticommons idea originated from Heller's broader theory derived from studying Russia's transition from communism to a market economy and received support from several economists.[56] However, despite the widespread diffusion of the anticommons theory, several shortcomings can be identified when it is applied to the patent system with regard to gene technologies. First, an implicit part of the anticommons argument is that there is a scarcity in the biological commons akin to a geographical scarcity.[57] However, the intangible commons in patent property rights differs from traditional tangible property like land, water, air, etc.[58] Furthermore, the 'geography' in genetic sciences is much more complex than physical land. For example in genetic research there can be many starting points and many routes that will lead to the desired final innovative product.[59] Moreover, unlike tangible properties, the intangible commons cannot be easily defined. It could be pictured as a cloud of information or ideas that, unlike tangible property, do not depreciate in the same way that tangible resources do when they are shared by multiple individuals.

Second, patent rights are limited in time. The anticommons tragedy in patent property rights would therefore be a temporary problem that could only have limited impact on future research. Due to the temporary nature of patents, patent holders will be able to enforce their rights of exclusion for a maximum of twenty years, and thus the negative externalities that

---

[55] Heller and Eisenberg (n 11); David E Adelman, 'A Fallacy of the Commons in Biotech Patent Policy' (2005) 20 BTLJ 985; David E Adelman, 'Reassessing the Anticommons Debate in Light of Biotechnology Patent Trends' in Peter K Yu (ed), *Intellectual Property and Information Wealth: Issues and Practices in the Digital Age* (Vol 2, Greenwood Publishing Group 2007) 302.

[56] James M Buchanan and Yong J Yoon, 'Symmetric Tragedies: Commons and Anticommons' (2000) 43 Journal of Law & Economics 1; Robert Cook-Deegan and Christopher Heaney, 'Patents in Genomics and Human Genetics' (2010) 11 Annual Review of Genomics and Human Genetics 383.

[57] Buchanan and Yoon (n 56); Frederic M Scherer, 'The Economics of Human Gene Patents' (2002) 77 Acad Med 1348.

[58] Samantha Leung, 'The Commons and Anticommons in Intellectual Property' [2011] UCL Jurisprudence Review 16.

[59] Ted Buckley, *The Myth of the Anticommons* (BIO 2007) 5.

result from an anticommons situation in the patent system would be deemed to disappear.

Third, the grant of exclusive patent rights does not necessarily imply that resources will be under exploited. Patent owners are given exclusion rights but they can also use their patents. Thus, they may exclude others from usage but at the same time, may directly use the facility or allow others access with permission.[60] Moreover, even during the period of patent protection, patent holders also face the possibility that new patents, old patents that have expired and thus become freely available, or new techniques that come into the public domain will erode their dominance.[61] Applicants usually file patent applications because they intend to make profits by utilizing the patent system. Thus, if patent holders do not practise their inventions, and nor do they license them, they would probably miss important income opportunities.

Fourth, patents help to facilitate investment and innovation in sectors where the activities are costly and risky. Therefore, the rights to restrict use that are granted to patent owners also serve to foster the development of certain industries.[62] For example, since patent protection is available, innovation in the biotech industry has progressed at high speed (see section 2.2.2).

Finally, there is a lack of exhaustive empirical evidence regarding the possible negative effects associated with patent clusters. Furthermore, in the case of gene patents the existence of contradictory studies concerning the implications of clusters in biotech innovation[63] poses difficulties for identifying a potential anticommons situation in this industry.

---

[60] Buchanan and Yoon (n 56).

[61] Richard A Epstein and Bruce N Kuhlik, 'Is There a Biomedical Anticommons' (2004) 27 Regulation 54.

[62] Scott Kieff, 'Facilitating Scientific Research: Intellectual Property Rights and the Norms of Science – A Response to Rai and Eisenberg' [2000] Law and Economics Papers, Working Paper 43. For example, a patent on a particular product may prevent other businesses from exactly copying that product, but businesses can learn the market value of the product by observing the patented product's success and this can spur them to try other related technologies, see Geradin, Layne-Farrar and Padilla (n 39).

[63] NIH, Report of the National Institutes of Health Working Group on Research Tools (June 1998); John P Walsh and Wesley M Cohen, 'Research Tool Patenting and Licensing and Biomedical Innovation, in Patents in the Knowledge-Based Economy' in Wesley M Cohen and Stephen Merrill (eds), *Patents in the Knowledge-Based Economy* (National Academies Press 2002). For an example of cases where the technology covered by research tool patents was broadly disseminated see Heather Hamme Ramirez, 'Defending the Privatization

Therefore, although the anticommons theory anticipates a possible consequence of excessive patenting in gene technologies, there is no substantial evidence to sustain such proposition. Nevertheless, even though the number of cases where patent monopolies have in reality hindered innovation is reduced, there is no guarantee that more cases will not arise in the future. For example, in the area of genetic research tools the increase in the number of patents over upstream innovations poses barriers to the development of important medical products such as diagnostic tests for personalized medicine.

7.3.3 The Case of Biotech Research Tools

The problems associated with extensive patenting are especially serious with regard to certain types of gene technologies where the proliferation of patent rights could stop or slow down the development of potentially life-saving products. This is the case for research tools, which are materials used in the laboratory to aid investigators in discovering and developing new products. They are sometimes referred to as 'upstream products' because they comprise early stage inventions that are used to develop final products. By contrast, end products developed through the use of upstream research tools are known as 'downstream inventions'.[64] Examples of research tools include:

- partial sequences of genes (ESTs)
- genes
- cell lines
- monoclonal antibodies
- reagents
- animal models
- growth factors
- combinatorial chemistry and DNA libraries
- clones and cloning tools
- methods
- laboratory equipment and
- machines.[65]

---

of Research Tools: An Examination of the Tragedy of the Anticommons in Biotechnology Research and Development' (2004) 53 Emory Law Journal 359.

[64] Hamme Ramirez (n 63).

[65] Michael S Mireles (n 6).

A frequently cited example of research tools are ESTs, which are very often used to find certain parts of DNA; ESTs can be used to identify an expressed gene and also as a sequence-tagged site marker to locate a particular gene on a physical map of a genome.[66] In contrast, examples of commercial applications of upstream research tools are gene therapies, diagnostic products and biologic drugs.

The distinction between research tools and downstream technologies is not clearly defined. Research tools may be developed by public entities, private companies or by entities that develop both upstream products and downstream commercial applications. They may also be included in the commercial applications sold to the end-user, like diagnostic tests. Moreover, companies can develop research tools themselves, or acquire them through purchasing the assets of another company or by obtaining a licence. In this regard, the cost involved in developing research tools in-house or acquiring the assets of a company can be very high, and thus, licensing is usually the most cost-effective method of obtaining the rights to use a research tool.[67] However, as explained above, licensing negotiations often involve structural problems (such as imbalanced bargaining positions) that could lead to an inefficient utilization of these technologies.

In human genetics, patent applications over research tools are frequently based on new discoveries where the utility can only be defined by reference to their value in performing further research. This situation may foster the growth of markets based on licences of patented upstream products and research discoveries that would have otherwise been in the public domain.[68] Since today most research tools are patentable, problems may arise when in order to develop new commercial technologies it is necessary to use multiple research tools each of which might require a separate licence. The royalties assigned to various licensors may severely erode profit potential, creating a disincentive for companies that require numerous research tools to develop specific commercial products or services.[69] Subsequently, the need to collect multiple rights upstream

---

[66] Cynthia D Lopez-Beverage, 'Should Congress Do Something about Upstream Clogging Caused by the Deficient Utility of Expressed Sequence Tag Patents' (2005) 10 J Tech L & Pol'y 35.

[67] Michael S Mireles (n 6).

[68] Roberto Mazzoneli and Richard R Nelson, 'The Benefits and Costs of Strong Patent Protection: A Contribution to the Current Debate' (1998) 27 Research Policy 273.

[69] Rebecca S Eisenberg, 'Why the Gene Patenting Controversy Persists' (2002) 77 Academic Medicine 1381; Michael S Mireles (n 6).

could have the effect of preventing the development of publicly beneficial products further downstream, which might in turn hinder future innovation in important areas such as biomedical research. This is for instance the case of biomarkers used in personalized medicine, which are usually subject to many different patents owned by different companies that, unless all of them are acquired, pose important barriers to the development of diagnostic test kits.

The case of extensive patenting in the area of research tools and its impact on subsequent innovation suggests that the possibility of an anticommons situation in gene technologies should not be completely disregarded. Moreover, it might be advisable to adopt measures to prevent the development of such blocking webs of patents from which inventors are compelled to obtain licences in order to be able to carry out further research.

7.4 RAISING THE INDUSTRIAL APPLICATION BAR FOR GENE PATENTS

As noted above, although empirical evidence regarding the existence of patent clusters in genetic research is not completely conclusive, past and present events suggest that extensive patenting in this and other fields might hinder access to important technological developments. In this regard, adopting adequate measures would help to prevent, or at least minimize, the potential negative effects of extensive patenting on research and innovation. However, the special characteristics of genetic inventions and the particular structural features of the sector (e.g. public–private nature, unclear distinction between upstream/basic and downstream/applied research, relatively new and fast-growing industry, and view of patents as essential means to foster innovation) make it hard to find a viable solution to the problem of patent conglomerates.

Some studies suggest that in several sectors patent protection is not the main mechanism through which firms appropriate the returns for their product and process innovations.[70] In those sectors, forbidding patents in order to impede the formation of patent clusters might be an option.

---

[70] See Richard C Levin and others, 'Appropriating the Returns from Industrial Research and Development' [1987] 3 Brookings Papers on Economic Activity (Special Issue on Microeconomics) 783; Wesley J Cohen, Richard R Nelson and John P Walsh, 'Protecting Their Intellectual Property Assets: Appropriability Conditions and Why the US Manufacturing Firms Patent (or Not)' [2000] NBER Working Paper No 7552.

However, innovation in life sciences industries is heavily dependent on patents, and thus, excluding biotech inventions from patent protection would not probably be the most effective policy mechanism.

The solutions most discussed in the patent law literature to address the issue of patent clusters include the so-called experimental use exception, cross-licensing, compulsory licensing and patent pools. In addition, there are also other measures that do not establish particular ex post mechanisms for facilitating access to patented technologies, but rely on ex ante policy measures that help control the formation of patent clusters. In this regard, policies to raise the bar in the application of patent law standards, particularly those provisions that relate to the rules on industrial application and scope of protection, have garnered increasing attention.

7.4.1 Experimental Use Exception

The experimental use exception is an exception to the rights of the patent proprietor, which allows third parties to use a patented invention for the purposes of research. This exception, which is especially relevant for the development of drugs, is part of European patent law. Although the EPC and the recently adopted regulation on the unitary patent make no explicit mention to it,[71] national patent laws generally include a provision stating that the rights conferred by a patent shall not extend to acts done for experimental purposes relating to the subject matter of the patented invention. However, interpretations of the scope and application of this exception vary across countries since there are usually considerable differences between the wording of the national legislations and interpretations of the exception by national courts.[72]

Moreover, in science-based industries like genetics, the line between commercial and non-commercial research is often difficult to draw; and thus, in order to avoid disparities, any research exception would need to

---

[71] See Regulation (EU) No 1257/2012 of the European Parliament and the Council of 17 December 2012 implementing enhanced cooperation in the area of the creation of unitary patent protection, OJ L361/1.

[72] See Geertrui van Overwalle and others, 'Models for Facilitating Access to Patents on Genetic Inventions' (2006) 7 Nat Rev Genet 143; Nicolas van Zeebroeck, Bruno van Pottelsberghe de la Potterie and Dominique Guellec, 'Patents and Academic Research: A State of the Art' (2008) 9 Journal of Intellectual Capital 246.

be carefully worded and implemented.[73] Today most research in biotechnology has commercial implications and patents are increasingly important for firms in this sector,[74] therefore the experimental use exception is unlikely to provide an optimal model for addressing access problems.

7.4.2 Licensing Agreements

The most employed mechanisms for gaining access to patented technologies are licensing agreements. A licence is a right granted by the patent holder to a third party to use the patented invention. Today there are a wide variety of licensing agreements, which offer flexible models that allow access and use conditions to be adjusted to specific needs and circumstances. For example, when two companies own complementary or blocking patents, cross-licensing helps to keep transaction costs to a minimum.[75] However, success in licensing negotiations usually depends on the perceived value of the patentee's patent portfolio. The stronger the patent, the stronger the bargaining position. Moreover, since licensing agreements are entirely voluntary, access to essential technologies is not guaranteed in all cases.[76] Therefore, although licensing agreements provide a simple and effective model for spreading the benefits of patented technologies, they entail important shortcomings such as the possible imbalances between the positions of the negotiating parties and the voluntary character of these types of contracts.

---

[73] Some authors consider that unless the exception is interpreted very narrowly, it could significantly erode patent protection in biotechnology and also deter investment in research and development, see David B Resnik, 'A Biotechnology Patent Pool: An Idea Whose Time Has Come?' (2003) 3 The Journal of Philosophy, Science & Law.

[74] Eisenberg 'Patents and the Progress of Science: Exclusive Rights and Experimental Use' (n 6).

[75] Van Overwalle and others (n 72).

[76] For an example, see Myriad Genetics' licensing approach regarding genes for the screening of breast cancer. The company only licensed the test exclusively to some commercial genetic laboratories that would market the tests within specific geographical regions and were only allowed to carry out testing of a limited set of breast cancer (BRCA) genes' mutations, while the complete sequence analysis is still carried out by Myriad, see E Richard Gold and Julia Carbone, 'Myriad Genetics: In the Eye of the Policy Storm' (2010) 12 Genetics in Medicine: official journal of the American College of Medical Genetics (Supp) S39. See also Esther van Zimmeren and Geertrui Van Overwalle, 'A Paper Tiger? Compulsory License Regimes for Public Health in Europe' (2011) 42 IIC 4.

7.4.3 Compulsory Licenses

Under the compulsory license mechanism a government or court can compel a patent holder to license his patent rights in order to meet different public purposes that go beyond the private interests of the patentee.[77] Article 31 of the TRIPS Agreement, which relates to non-voluntary licences granted to third parties, affirms the right of Member States to grant compulsory licences and gives them autonomy to determine the grounds on which such licences can be granted.[78] Today virtually all countries around the world allow compulsory licences in their national legislation, which can be granted either by a judge or the government.[79]

It has been suggested that the compulsory licensing mechanism could be invoked to address the potential negative effects of patent clusters in biotechnology.[80] However, this solution is of exceptional character and thus it can only be used under extraordinary circumstances, such as:

- when a demand for the patented product is not being met
- there is a public interest in the patented invention
- the proprietor of the patent refuses to grant a licence or licences on reasonable terms or
- the patentee has imposed conditions on the grant of licences that unfairly prejudice the manufacture of the invention, the use of materials not protected by the patent or the establishment of industrial activities.

Nonetheless, following the controversy surrounding patents related to diagnostic genetic testing, specially tailored compulsory licence mechanisms with the aim of serving public health have been established in France, Belgium and Switzerland.[81]

---

[77] Duncan Matthews, *Globalising Intellectual Property Rights* (Routledge 2002) 77.

[78] TRIPS Agreement (n 20) art 31.

[79] See also Van Zimmeren and Van Overwalle (n 76).

[80] Van Overwalle and others (n 72).

[81] For example Article 40(b) of the Swiss Patent Act of 1954 contains a specific provision for the non-voluntary and non-exclusive licensing of research tools aimed at ensuring access to essential upstream discoveries such as gene sequences and methods such as polymerase chain reaction (PCR). Besides, Articles L613–616 of the French Intellectual Property Code of 1992 and Article 31bis of the Belgium Patent Law of 1984 contain similar provisions with regard to ex-officio licences for public health to genetic diagnostics.

7.4.4 Patent Pools

Even though existing evidence does not entirely support claims that patent pools encourage innovation,[82] so far, the establishment of patent pools is considered to be the most promising mechanism to facilitate access to patent clusters. Patent pools are agreements between two or more patent owners to license one or more of their patents to one another, or to license them as a package (one-stop shop) to third parties that are willing to pay the royalty fee associated with the licence. Those licences are provided to the licensee either directly by the patent holders or indirectly through a new entity that is specifically created for administering the pool.[83] In technical fields other than genetics, patent pools have emerged to deal with the problems associated with patent clusters. For example, in 1917 an aircraft pool was formed that encompassed almost all aircraft manufacturers at the time the US entered World War I.[84] In the late 1990s, several patent pools were also formed in the electronics and telecommunications industries, for instance, the moving picture experts group (MPEG)-2 pool in 1997 for inventions relating to the MPEG-2 standard.[85] Patent pools can contribute to:

- limiting royalty stacking and transaction costs
- reducing patent litigation
- institutionalizing information exchange
- allowing and encouraging access to the pooled technologies and
- spreading the risks and benefits of technology implementation among the participants of the pool.[86]

---

[82] Ryan L Lampe and Petra Moser, 'Do Patent Pools Encourage Innovation? Evidence from the 19th-Century Sewing Machine Industry' (2009) National Bureau of Economic Research, Working Paper 15061.

[83] Andewelt (n 35); Van Overwalle and others (n 72).

[84] The Manufacturers Aircraft Association (MAA), formed on 12 July 1917, combined all patents that were needed to build a plane and made them available for licensing, see Lampe and Moser (n 82).

[85] Verbeure and others (n 42).

[86] Frank Grassler and Mary Ann Capria, 'Patent Pooling: Uncorking a Technology Transfer Bottleneck and Creating Value in the Biomedical Research Field' (2003) 9 Journal of Commercial Biotechnology 111.

Furthermore, patent pools have often been a successful solution to conflicts over standards, particularly when each firm's patents only cover a small component of a product.[87]

In the electronics and telecommunications sector, the generation of internationally accepted technical standards is seen as a strong incentive for setting up patent pools.[88] However, such a standard is missing in genetics.[89] Therefore, in the absence of this type of standard-driven incentive, dominant players in the biotech industry might be reluctant to join a pool because there is no apparent gain.[90] Moreover, since biotech firms used to rely on their patent portfolio for securing funding and negotiating licensing agreements, they might refuse to share their intellectual property assets.

With regard to antitrust issues, patent pools may be found to be pro-competitive if they:

- integrate complementary technologies that are 'essential' to a standard
- clear blocking positions
- reduce transaction costs
- avoid costly infringement litigation and
- promote the dissemination of technology.[91]

However, pools may be found to be anti-competitive if they constitute methods of fixing prices or allocating customers and markets, exclude competitors or discourage participants and third parties from engaging in research and development.[92] The biotech industry, particularly the area of

---

[87] Gallini (n 1). Standards are technical specifications relating to a product or an operation, which are recognized by a large number of manufacturers and users. Standards can be an important trigger to set up a pool, as illustrated in the electronics and telecommunications sectors, and this might also be true in the field of genetics, see Verbeure and others (n 42).

[88] See Van Overwalle and others (n 72); Verbeure and others (n 42).

[89] Van Overwalle and others (n 72).

[90] Verbeure and others (n 42). For some examples about possible means for creating patent pools in biotechnology, see Grassler and Capria (n 86); Resnik (n 73); T J Ebersole, C Guthrie and J A Goldstein, 'Patent Pools and Standard Setting in Diagnostic Testing' (2005) 23 Nat Biotech 937; Van Overwalle and others (n 72).

[91] Andewelt (n 35).

[92] In the European Union, the major competition laws relating to technology licensing are laid down in the Regulation (EC) No 772/2004 of 7 April 2004 on the application of Article 81(3) of the Treaty to categories of technology transfer

genetics, is dispersed, does not have common goals and advances too quickly, making it difficult to identify the essential patents for a pool especially in early stage biotechnology areas.[93] In this sense, difficulties in defining technical standards and thus identifying which complementary patents are essential to the pool may complicate and delay proceedings with antitrust authorities.

Nonetheless, there are some cases of successful patent pools in gene technologies. The Golden Rice pool is an example of how private and public organizations, in a combined effort, dealt with surrounding patents to create a non-profit humanitarian patent pool in the form of a single licensing authority.[94] In this case, Potrykus succeeded in genetically enriching rice grains with β-carotene24 (Golden Rice) and wanted to transfer the materials to developing countries for further breeding in order to introduce the trait into local varieties that are consumed in these countries. Six key patent holders were approached and an agreement was reached that allowed Potrykus to grant licences free of charge to developing countries, with the right to sub-license. Subsequently, a humanitarian board was established as a voluntary association to assist in the associated governance and decision-making.[95] A similar example is the Severe Acute Respiratory Syndrome (SARS) corona virus pool, which is supported by the World Health Organization (WHO).[96] However, these examples are very atypical and are based on altruistic purposes that most biotechnology companies do not share.

With features similar to patent pools, 'open source' models where participation is open for all in exchange for a fair price and certain conditions have recently been implemented successfully.[97] Nonetheless,

---

agreements, OJ L 123/11 and the Guidelines on the application of Article 81 of the EC Treaty to technology transfer agreements [2004] OJ C 101/2. See Gallini (n 1); Ebersole, Guthrie and Goldstein (n 90). See also Joshua A Newberg, 'Antitrust, Patent Pools, and the Management of Uncertainty' (2000) 3 Atlantic Law Journal 1.

[93] Ebersole, Guthrie and Goldstein (n 90).

[94] See Van Overwalle and others (n 72).

[95] Verbeure and others (n 42).

[96] James HM Simon and others, 'Managing Severe Acute Respiratory Syndrome (SARS) Intellectual Property Rights: The Possible Role of Patent Pooling' (2005) 83 Bulletin of the World Health Organization 707.

[97] See for example Syngenta AG's e-licensing platform named TraitAbility, which provides 'quick and easy' access to Syngenta's patents on native traits of commercial vegetable varieties with transparent and FRAND (fair, reasonable and non-discriminatory) conditions (Syngenta AG's e-licensing platform TraitAbility <http://www3.syngenta.com/global/e-licensing/en/e-licensing/About/Pages/

both patent pools and open innovation models rely on the voluntary engagement of the patentees, and thus, they do not offer a valid solution in cases where patent holders refuse to licence their technologies. Furthermore, the examples above relate to the agricultural field, where companies have different characteristics and objectives than those in the pharmaceutical sector.

Institutions like the OECD consider the concept of a patent pool to be an interesting one for biotechnology but have some doubts as to whether the technologies and markets for genetic inventions are amenable to patent pools.[98] In December 2000 the USPTO also distributed a white paper that developed the idea of a patent pool for biotechnology. The paper outlined some of the benefits and risks associated with patent pooling, as well as some legal restrictions. It concluded that pooling is a 'win–win' situation for the interests of both public and private industries.[99] However, a biotechnology patent pool would only be effective if the right balance between the cost of creating a pool and the prospect of adequate revenue generated by royalties for the end product is achieved.[100] Thus, an important obstacle to starting, developing and sustaining a patent pool is convincing the various parties that there is substantial economic benefit.[101] Therefore, although in theory patent pools might seem to be a good solution to the potential problems related to the proliferation of patent clusters, when applied to biotech industries several difficulties arise such as a lack of technical standards, antitrust issues and limited and specific examples in gene-related technologies. Moreover, the important role of patents in biotech industries makes it difficult to build a patent pool based on the volunteer motivation of participants.

About.aspx> accessed 20 February 2015). See also the International Licensing Platform (ILP) aimed at improving access to and use of plant breeding traits for vegetables that are protected by patents or plant breeders' rights.

[98] OECD, 'Genetic Inventions, Intellectual Property Rights and Licensing Practices, Report of a workshop organized by the OECD Working Party on Biotechnology' (n 4) 67.

[99] USPTO, 'Patent Pools: A Solution to the Problem of Access in Biotechnology Patents?' (USPTO 2000); Resnik (n 73).

[100] USPTO, 'Patent Pools: A Solution to the Problem of Access in Biotechnology Patents?' (n 99); Resnik (n 73); Verbeure and others (n 42).

[101] Even then, there would probably be some holdouts. For example, a company with patents related to a valuable protein, for example erythropoietin, would probably not place this patent in the pool because it would find it more profitable licensing it separately than cooperating with other patent holders, see Resnik (n 73).

7.4.5 Raising the Industrial Applicability Bar

The models explained above could in some cases offer suitable solutions to the problematic effects of extensive patenting by improving access to the patented technologies forming the patent cluster. However, these mechanisms are insufficient in the sense that they do not prevent the formation of such conglomerates, but can only reduce the impact of their negative effects by providing access to the patented innovations on limited occasions. In this regard, adopting adequate policy and legal measures would help to prevent the excessive growth and fragmentation of patent rights in gene technologies.

For instance, it has been proposed that patent standards should be strengthened so as to make sure that no patents are granted on pure scientific discoveries without industrial application or utility, and that patents are only granted to new and non-obvious inventions.[102] Doing so would also allow a line to be drawn between science and technology and would probably avoid some of the most potentially threatening patents, such as patents on mere discoveries of human DNA sequences, to be granted or enforced.[103] In this regard, the Biotech Directive's approach towards the interpretation of the industrial application of gene patents would help to prevent massive patenting in an emerging and fast-evolving industry like human genetics, where dependency on upstream patents without a known practical function may pose serious barriers to the progress of future research.

The most immediate consequence of the boom in patent applications over DNA inventions is a general increase in the workload of the EPO and all national patent offices. In this sense, low patenting standards, especially for new technologies, together with an increase in the workload of patent offices may have serious consequences for the quality of issued patents. Besides, in gene technologies the speed at which new subject matter and scientific inventions are introduced in the patent system makes it harder to assess the state of the art, and thus, to determine whether the claimed invention is novel, involves an inventive

---

[102] Van Zeebroeck, Van Pottelsberghe de la Potterie and Guellec (n 72).

[103] Ibid. See also Andrew Gowers, *Gowers Review of Intellectual Property* (Independent report, UK Government 2006) 32.

step and is industrially applicable.[104] In this sense, although it is difficult to document, the quality of patents is reported to be declining.[105]

As explained in the previous chapters, with regard to human genetic inventions, Article 5(3) of the Biotech Directive significantly facilitates a rigorous assessment of the industrial application requirement by imposing on applicants the obligation of disclosing the invention's utility already present in their patent applications. Moreover, after the introduction of the Biotech Directive, the extent of the scope of protection of this type of invention is defined according to the practical utility that the patentee claims. Therefore, raising the industrial application bar in order to prevent the negative effects of patent clusters in gene technologies is an approach fully supported by current European patent law and policy with regard to inventions over human genetic material. Furthermore, as explained in Chapter 6, the effects of the Biotech Directive's approach towards the industrial application of human DNA inventions may extend to inventions in other fields, thus allowing the possibility of using this requirement to address the issue of patent conglomerates in other technology areas.

Patent law should not be seen as a goal itself but as a means for achieving objectives of innovation and technological development.[106] Therefore, interpreting existing legislation in light of the realities of biotechnology and modern patent practice would certainly enhance the role of patent laws in life sciences innovation. In this sense, without amending European patent law, the Biotech Directive standard of industrial application could restrict the emergence of patent floods and subsequent barriers to innovators. Precautionary measures, such as raising the bar of industrial application, would help to maximize the scientific, technological and social benefits of patenting while minimizing the risk of excessive patent conglomerates and thus maintaining the qui pro quo of patent systems.[107] Moreover, in light of Article 9 of the Biotech Directive, this solution would also serve to limit the scope of such patents to the actual function of the protected inventions, hence

---

[104] Cowin and others, 'Policy Options for the Improvement of the European Patent System' (n 31) 20.

[105] It has been reported that there is an increase in the number of bad quality patents, trivial patents or patents for insignificant inventions, see Cowin and others, 'Policy Options for the Improvement of the European Patent System' (n 31) 20.

[106] See Mark A Lemley, 'Software Patents and the Return of Functional Claiming' [2012] Stanford Public Law Working Paper No 2117302.

[107] See Resnik (n 73).

allowing inventors to access the patented technologies for any other uses falling outside the patent's precise scope of protection.

This approach does not aim to change patent laws or encourage the adoption of new legislation. On the contrary, it offers an ad hoc solution to prevent the possible formation of detrimental patent clusters that can be incorporated into the current system without modifying existing rules. In this regard, a 2005 report of the European Parliament encouraged the adjustment of the system as it now stands with proposals that further the objectives to stimulate innovation and the diffusion of knowledge.[108] Therefore, developing a strict approach towards the interpretation of the requirement of industrial application as an ex ante measure to restrict the emergence of patent clusters in gene technologies would also be in accordance with such a suggestion.

7.5 CONCLUDING REMARKS

Due to the increasing levels of patenting activity in genetic industries, the idea of a possible cluster of fragmented patent rights that may restrict access to essential technologies, and thus create an anticommons situation that could hinder future innovation, has garnered special attention. This issue has been occasionally discussed in the patent law literature regarding biotechnology. However, there is a lack of substantial empirical evidence on the existence and potential negative effects of patent floods in this field. Nevertheless, recent cases, along with the importance of life sciences innovation for societies worldwide, invite policy makers and stakeholders in general to consider this issue as more than a simple theoretical possibility.

In recent years, several measures to reduce the negative effects associated with patent clusters have been proposed. In particular, cross-licensing, patent pools and open innovation models are considered to be the most promising solutions to the accumulation of interdependent patents. However, they present important limitations when applied to human gene technologies. For example, the lack of technology standards and the voluntary character of these agreements make it difficult for biotechnology companies, which depend so much on patent monetization, to engage in such negotiations. In this regard, it is argued that the current European approach towards the industrial application of genetic inventions in its different dimensions – by reducing the patentability of

[108] Cowin and others, 'Policy Options for the Improvement of the European Patent System' (n 31) 20.

discoveries, raising the industrial application standard and restricting the scope of gene patents – can help to avoid the emergence of unduly restricting patent conglomerates. A strict interpretation of this requirement can reduce the number of patents over discoveries with no practical utility, and also the amount of patents the scopes of which do not correspond with the technical utility claimed by the applicant. Furthermore, this policy option does not require modifying existing patent laws since the Biotech Directive and the EPC already provide a valuable platform for implementing such an approach.

Thus, in addition to encouraging patentees to engage in ex post licensing mechanisms to improve access to patented technologies, patent authorities can interpret the requirement of industrial application with a view towards avoiding excessive patenting and fragmentation of rights in specially problematic areas like genetic diagnosis where extensive patenting can pose serious barriers to the progress of research in this prominent field.

8. Conclusion

Applying patent law rules adequately is crucial for meeting the objectives of fair reward that the patent system pursues. In the case of human gene inventions, the requirement of industrial application is particularly important since it ensures that exclusive rights over human body parts such as sequences or partial sequences of human genes are granted only to those creations that bring a real technical contribution to society. Given the importance of human gene inventions, especially in healthcare, it is crucial that society is only prevented from freely using inventions that meet such criterion.

Concerns about the patentability of human DNA sequences before a specific function of the claimed genetic sequence has been identified have been part of the international debate over patenting biotechnological inventions since the 1980s. Patent rights are very important for innovation in biotechnology industries; however, allowing patents over human gene parts without a known application in industry would in the end lead to an imbalance in the patent bargain in which exclusive rights are granted to inventors in exchange for the disclosure of innovative technologies that are useful for society.

The 1998 Biotech Directive's policy approach of requiring applicants to disclose the industrial application of the claimed human gene sequence has been able to address such concerns efficiently. There are important arguments against the adoption of a strict interpretation of the requirement of industrial application. With regard to human DNA patents, it has been claimed that the commercial value of isolated human genetic material arises prior to and goes beyond the identification of a specific utility in industry, and that preventing companies from patenting gene fragments at an early stage would affect the capacity of firms to recover the invested resources and thus interfere in the progress of genetic research. Besides, arguments have been raised regarding the discriminatory conditions that adopting stricter patentability standards for particular technologies may impose on inventors in such sectors.

However, as discussed in Chapters 3 and 4, a system that applies a very loose industrial application criteria would arguably serve the objectives of fair reward and dissemination of knowledge that justify the granting of patent rights. In this regard, it is argued that the Biotech Directive's strict approach towards the industrial applicability of human gene inventions achieves an optimal balance between the private rights of the patentee and the public interest. It has been shown that this interpretation serves to impede the grant of patents over vague and speculative claims that require further research to identify any practical utility. Furthermore, the EPO has pragmatically implemented this approach on a case-by-case basis that takes into consideration the particular characteristics that genetic inventions may present, for example with regard to the type of function that should be disclosed, the admissibility of results obtained through computer-assisted methods or the need to disclose a specific industrial application when the claimed invention is essential for human health purposes.

Besides, the analysis of European patent law and policy with regard to the industrial applicability of human genetic material is also important for understanding the interpretation of this requirement for non-human genetic material. For instance, the EPO's interpretation of the Biotech Directive's function disclosure requirement for human DNA sequences has become part of the guidelines for interpreting the general requirement of industrial applicability. Moreover, since the US has taken a similar approach towards the utility of genetic and non-genetic inventions,[1] such analysis also assists with understanding US patent policy on the requirement of utility and its implications for patent applicants.

Further, the Biotech Directive's industrial application disclosure requirement for human DNA sequences also plays a key role in the distinction between discoveries and inventions in this field. As analysed in Chapter 5, the exclusion from patentability of discoveries in European patent law is especially relevant in the case of inventions concerning naturally occurring genes, in the sense that discoveries of natural substances must acquire a technical character that goes beyond the abstract characteristics of living matter in its natural state. The EPO has traditionally solved this issue through the 'isolation doctrine', which allows the patentability of natural substances as long as they are isolated from their original surrounding. However, with the adoption of the Biotech Directive, in addition to isolating the human genetic material, applicants must specify the utility of the newly discovered sequence in the patent application in order for the claimed human DNA substance to acquire an invention's technical character. This interpretation of the

[1] United States Patent and Trademark Office (USPTO), Utility Examination Guidelines (2001).

exclusion of discoveries is consistent with traditional views of the European patent law requirement for an invention that requires discoveries to be reduced to useful creations. Moreover, this double requirement further defines the distinction between discoveries and inventions in this context in view of the fact that all genes that are discovered have to be isolated by technical means first; which makes them accessible for further study but not for practical exploitation if the function of the isolated sequence is not determined.

In addition, the criterion of industrial application has become the reference point for determining the scope of protection of all types of gene patents. Article 9 of the Biotech Directive extends the scope of protection of genetic patents to the ability of the claimed invention to perform its function. This provision has been interpreted by the majority as an open gate for establishing purpose-bound protection for this type of invention, and the Parliament, the CJEU and several national authorities have acknowledged that this article indeed abolishes absolute product protection in favour of purpose-bound protection limited to the industrial application disclosed by the applicant for all inventions concerning genetic information.

Such interpretation is in line with those economic theories supporting narrow scopes of protection in very innovative scientific fields like biotechnology where unduly broad scopes of protection may seriously undermine the progress of further research. In particular, restricting the scope of protection of human genetic inventions to their capability of carrying out their intended utility helps to strike the balance between patent holders and third parties, and promote subsequent innovation by ensuring that patented genes are freely available for exploitation and further research as long as the protected material does not perform the specific function claimed by the patentee. Further, interpreting that function-related scope of protection of gene patents in general, human and non-human, by making reference to the Biotech Directive's requirement of disclosing the industrial application of human gene sequences, would imply that the latter provision becomes applicable to all types of genetic inventions and not limited to human DNA. Therefore, the implications of adopting a strict approach towards the standard of industrial applicability would ultimately extend to all sorts of genetic inventions.

Besides this, the policy approach taken by the Biotech Directive towards the industrial applicability standard in genetic patents, which has as a matter of fact impacted on the decisions of the EPO and domestic courts, coupled with the CJEU's decision in Monsanto, represents another step forward in the increasing involvement of the EU in

European patent law and intellectual property law in general. Moreover, should the EU unitary patent package finally enter into force, it can be expected that the Unified Patent Court (UPC) will follow, and further settle, the same approach towards the industrial applicability and scope of protection of genetic patents, since the UPC's jurisprudence will necessarily be based on the sources of law indicated in Article 24 of the UPC Agreement;[2] that is:

- EU law
- the UPC Agreement
- the EPC
- international agreements applicable to patents and binding on all the Contracting Member States and
- national laws, having EU laws primacy over the other legal sources under Article 20 of the UPC Agreement.

Finally, this book acknowledges the serious consequences that extensive patenting and fragmentation of patent rights may have for research and innovation in gene technologies. Within this context, it is argued that the Biotech Directive's policy initiative of raising the industrial applicability bar can act as an ex ante mechanism for reducing the patenting of overly broad patents on genetic discoveries that would unduly contribute to the formation of clusters of overlapping patents that pose barriers to future innovators. Besides, although this policy option does not provide a comprehensive solution to the problems associated with patent clusters, it can be an efficient prior and complementary means to other promising ex post models for facilitating access to important gene patents such as biomedical technologies. In this sense, the requirement of industrial application would be part of a set of available measures for preventing the negative effects of extensive patenting of human DNA sequences.

In sum, the European strict approach towards the industrial application of human gene patents has finally resulted in a system that offers a balanced and compromised policy option that is able to address key questions and objectives with regard to gene patents. In this regard, the book demonstrates that this approach towards the patentability of human genes is consistent with the origin and rationale behind the European requirement of industrial application, as well as the economic justification for the granting of patent rights as a means to promote innovation and dissemination of useful technologies.

---

[2] Agreement on a Unified Patent Court of 20 June 2013 [2013] OJ C 175/01.

Furthermore, it appears that adopting a strict interpretation of existing patent law standards in order to address the particular concerns that new technologies raise is certainly feasible. The Biotech Directive's policy on raising the industrial application bar within the context of human gene patents has been effectively implemented by the EPO and is progressively being introduced in national patent systems. Moreover, this has been achieved despite the complex characteristics of human gene technologies. Thus, the Biotech Directive's approach can serve as an example for dealing with future cases where challenging new technologies pose difficulties to the patent system and the introduction of specific criteria may be required. The case of the requirement of industrial application in human gene inventions suggests that modifying the interpretation of patent law standards to deal with the particular challenges that new technologies may pose is possible and can be done efficiently across European countries, regardless of the complexities that a given scientific field may present.

Bibliography

BOOKS

Bakels, R, *The Technology Criterion in Patent Law: A Controversial but Indispensable Requirement* (Wolf Legal Publishers 2012).

Beetz, R, Dieter Behrens and Wolfgang Dost, *European Patent Law: Practicing under the European Patent Convention (EPC)* (Heymanns 1979).

Beier, FK, Stephen Crespi and Joseph Straus, *Biotechnology and Patent Protection: An International Review* (OECD Publishing 1985).

Bliss, M, *The Discovery of Insulin* (Macmillan Press 1982).

Bostyn, SJR, *Enabling Biotechnological Inventions in Europe and the United States: A Study of the Patentability of Proteins and DNA Sequences with Special Emphasis on the Disclosure Requirement* (European Patent Office 2001).

Bovenberg, JA, *Property Rights in Blood, Genes and Data: Naturally Yours?* (Martinus Nijhoff Publishers 2006).

Buckley, T, *The Myth of the Anticommons* (BIO 2007).

Bud, R, *The Uses of Life: A History of Biotechnology* (Cambridge University Press 1993).

Burchfiel, KJ, *Biotechnology and the Federal Circuit* (BNA Books 1995).

Casper, S, *Creating Silicon Valley in Europe: Public Policy towards New Technology Industries* (Oxford University Press 2007).

Chisum, DS and others (eds), *Principles of Patent Law: Cases and Materials* (3rd edn, Foundation Press 2004).

Coke, E, *The Third Part of the Institutes of the Law of England: Concerning High Treason, and Other Pleas of the Crown, and Criminal Causes* (4th edn, first published by Andrew Crooke and others in 1669).

Cook, T, *Pharmaceuticals, Biotechnology and the Law* (2nd edn, Lexis Nexis 2009).

Cook, T, *A User's Guide to Patents* (A&C Black 2011).

Cornish, WR, David Llewelyn and Tania Aplin, *Intellectual Property: Patents, Copyright, Trade Marks & Allied Rights* (7th edn, Sweet & Maxwell 2010).

Crespi, RS, *Patenting in the Biological Sciences: A Practical Guide for Research Scientists in Biotechnology and the Pharmaceutical and Agrochemical Industries* (Wiley 1982).

Drahos, P, *A Philosophy of Intellectual Property* (Dartmouth 1996).

Dutfield, G, *Intellectual Property Rights and the Life Science Industries: A 20th Century History* (Ashgate 2003).

Fisher, M, *Fundamentals of Patent Law: Interpretation and Scope of Protection* (Hart Pub 2007).

Fysh, M and others, *The Modern Law of Patents* (2nd edn, LexisNexis Butterworths 2010).

Gold, ER, *Body Parts: Property Rights and the Ownership of Human Biological Materials* (Georgetown University Press 1996).

Grandstrand, O, *The Economics and Management of Intellectual Property: Towards Intellectual Capitalism* (Edward Elgar 1999).

Grubb, PW and Peter R Thomsen, *Patents for Chemicals, Pharmaceuticals, and Biotechnology: Fundamentals of Global Law, Practice, and Strategy* (Oxford University Press 2010).

Guellec, D and Bruno van Pottelsberghe de la Potterie, *The Economics of the European Patent System: IP Policy for Innovation and Competition* (Oxford University Press 2007).

Haedicke, MW and Henrik Timmann (eds), *Patent Law: A Handbook on European and German Patent Law* (CH Beck, Hart and Nomos 2014).

Heller, MA, *The Gridlock Economy: How Too Much Ownership Wrecks Markets, Stops Innovation, and Costs Lives* (Basic Books 2010).

Kamstra, G and others, *Patents on Biotechnological Inventions: The EC Directive* (Sweet & Maxwell 2003).

Kaufer, E, *The Economics of the Patent System* (Routledge 2001).

Lodish, HF and others (eds), *Molecular Cell Biology* (7th edn, WH Freeman and Co 2013).

MacQueen, HL, *Contemporary Intellectual Property: Law and Policy* (2nd edn, Oxford University Press 2011).

Matthews, D, *Globalising Intellectual Property Rights* (Routledge 2002).

Merges, RP, *Justifying Intellectual Property* (Harvard University Press 2011).

Mills, O, *Biotechnological Inventions, Moral Restraints & Patent Law* (Revised Edition, Ashgate 2010).

Nordhaus, W, *Invention, Growth and Welfare: A Theoretical Treatment of Technological Change* (Cambridge, MA: MIT Press 1969).

Pila, J, *The Requirement for an Invention in Patent Law* (Oxford University Press 2010).

Pires de Carvalho, N, *The TRIPS Regime of Patent Rights* (Kluwer Law International 2002).

Pires de Carvalho, N, *The TRIPS Regime of Patent Rights* (3rd edn, Kluwer Law International 2010).

Pisano, GP, *Science Business: The Promise, the Reality, and the Future of Biotech* (Harvard Business School Press 2006).

Pottage, A and Brad Sherman, *Figures of Invention: A History of Modern Patent Law* (Oxford University Press 2010).

Rimmer, M, *Intellectual Property and Biotechnology: Biological Inventions* (Edward Elgar 2008).

Sasson, A, *Medical Biotechnology: Achievements, Prospects and Perceptions* (UN University Press 2005).

Schumpeter, J, *Capitalism, Socialism and Democracy* (5th edn transferred to digital printing, Routledge 2005).

Scotchmer, S, *Innovation and Incentives* (MIT Press 2004).

Smith, A, *An Inquiry into the Nature and Causes of the Wealth of Nations* (edited by Edwin Cannan in 1904, first published by W Strahan and T Cadell in 1776).

Sommer, T, *Can Law Make Life (too) Simple: From Gene Patents to the Patenting of Environmentally Sound Technologies* (1st edn, DJØF Publishing 2013).

Sreenivasulu, NS and CB Raju, *Biotechnology and Patent Law: Patenting Living Beings* (Manupatra 2008).

Stack, AJ, *International Patent Law: Cooperation, Harmonization, and an Institutional Analysis of WIPO and the WTO* (Edward Elgar 2011).

Sterckx, S and Julian Cockbain, *Exclusions from Patentability: How Far Has the European Patent Office Eroded Boundaries?* (Cambridge University Press 2012).

Taylor, CT and Z Aubrey Silberston, *The Economic Impact of the Patent System: A Study of the British Experience* (Cambridge University Press 1973).

Tropp, BE, *Molecular Biology: Genes to Proteins* (4th edn, Jones & Bartlett Learning 2012).

Tudge, C, *In Mendel's Footnotes: An Introduction to the Science and Technologies of the Nineteenth Century to the Twenty-Second* (Jonathan Cape 2000).

Van Caenegem, W, *Technology Law and Innovation* (Cambridge University Press 2007).

Vettel, E, *Biotech: The Countercultural Origins of an Industry* (University of Pennsylvania Press 2006).

Warren-Jones, A, *Patenting rDNA: Human and Animal Biotechnology in the United Kingdom and Europe* (Lawtext Pub 2001).

Watson, JD, *The Double Helix: A Personal Account of the Discovery of the Structure of DNA* (1st Atheneum paperback edn, Atheneum 1980).

Watson, JD (ed), *Molecular Biology of the Gene* (7th edn, Pearson 2014).

Wegner, HC, *Patent Harmonization* (Sweet & Maxwell 1993).
Westerlund, L, *Biotech Patents: Equivalence and Exclusions under European and U.S. Patent Law* (Kluwer Law International 2002).
Wilson, K and John M Walker (eds), *Principles and Techniques of Biochemistry and Molecular Biology* (7th edn, Cambridge University Press 2009).

BOOK CHAPTERS

Acharya, R, Anthony Arundel and Luigi Orsenigo, 'The Evolution of European Biotechnology and Its Future Competitiveness' in Jaqueline Senker (ed), *Biotechnology and Competitive Advantage* (Edward Elgar 1998).
Adelman, DE, 'Reassessing the Anticommons Debate in Light of Biotechnology Patent Trends' in Peter K Yu (ed), *Intellectual Property and Information Wealth: Issues and Practices in the Digital Age* (Vol 2, Greenwood Publishing Group 2007).
Arrow, KJ, 'Economic Welfare and the Allocation of Resources for Inventions' in Richard R Nelson (ed), *The Rate and Direction of Inventive Activity: Economic and Social Factors* (Princeton University Press 1962).
Asheim, B, Finn Valentin and Christian Zeller, 'Intellectual Property Rights and Innovation Systems: Issues for Governance in a Global Context' in David Castle (ed), *The Role of Intellectual Property Rights in Biotechnology Innovation* (Edward Elgar 2009).
Boadi, RY, 'The Role of IPRs in Biotechnology Innovation: National and International Comparisons' in David Castle (ed), *The Role of Intellectual Property Rights in Biotechnology Innovation* (Edward Elgar 2009).
Bostyn, SJR, 'A Decade After the Birth of the Biotech Directive: Was It Worth the Trouble?' in Emanuela Arezzo and Gustavo Ghidini (eds), *Biotechnology and Software Patent Law* (Edward Elgar 2011).
Burk, DL and Mark A Lemley, 'Tailoring Patents to Different Industries' in Emanuela Arezzo and Gustavo Ghidini (eds), *Biotechnology and Software Patent Law: A Comparative Review of New Developments* (Edward Elgar 2011).
Cockburn, IM, 'State Street Meets the Human Genome Project: Intellectual Property and Bioinformatics' in Robert W Hahn (ed), *Intellectual Property Rights in Frontier Industries: Software and Biotechnology* (AEI-Brookings Joint Center for Regulatory Studies 2005).

Frahm, K and Sture Rygaard, 'An Introduction to European Intellectual Property Rights in Intellectual Property' in Paul England (ed), *Intellectual Property in the Life Sciences: a Global Guide to Rights and their Applications* (Globe Law and Business 2011).

Hahn, RW, 'An Overview of the Economics of Intellectual Property Protection' in Robert W Hahn (ed), *Intellectual Property Rights in Frontier Industries: Software and Biotechnology* (AEI-Brookings Joint Center for Regulatory Studies 2005).

Kieff, FS, 'On the Economics of Patent Law and Policy' in Toshiko Takenaka (ed), *Patent Law and Theory* (Edward Elgar 2008).

Macchia, G, 'Patentability Requirements of Biotech Inventions at the European Patent Office: Ethical Issues' in Roberto Bin and others (eds), *Biotech Innovations and Fundamental Rights* (Springer 2012).

May, C, 'On the Border: Biotechnology, the Scope of Intellectual Property and the Dissemination of Scientific Benefits' in David Castle (ed), *The Role of Intellectual Property Rights in Biotechnology Innovation* (Edward Elgar 2009).

Merges, RP and Richard R Nelson, 'Market Structure and Technical Advance: The Role of Patent Scope Decisions' in Thomas M Jorde and David J Teece (ed), *Antitrust, Innovation and Competitiveness* (Oxford University Press 1992).

Pagenberg, J and Uta Köster, 'History of Article 69 EPC' in Jochen Pagenberg and William R Cornish (eds), *Interpretation of Patents in Europe: Application of Article 69 EPC* (Heymanns 2006).

Pila, J, 'The Future of the Requirement for an Invention: Inherent Patentability as a Pre- and Post-Patent Determinant' in Emanuela Arezzo and Gustavo Ghidini (eds), *Biotechnology and Software Patent Law* (Edward Elgar Publishing 2011).

Prinz zu Waldeck und Pyrmont, W, 'Special Legislation for Genetic Inventions – A Violation of Article 27(1) TRIPS?' in Wolrad Prinz zu Waldeck und Pyrmont and other (eds), *Patents and Technological Progress in a Globalized World: Liber Amicorum Joseph Straus* (Springer 2009).

Schertenleib, D, 'An Introduction to European Intellectual Property Rights in Intellectual Property' in Paul England (ed), *Intellectual Property in the Life Sciences: a Global Guide to Rights and their Applications* (Globe Law and Business 2011).

Scott, J, 'Rational Choice Theory' in Gary Browning, Abigail Halcli and Frank Webster (eds), *Understanding Contemporary Society: Theories of the Present* (SAGE 2000).

Shapiro, C, 'Navigating the Patent Thicket: Cross Licenses, Patent Pools, and Standard-Setting' in Adam B Jaffe, Josh Lerner and Scott Stern (eds), *Innovation Policy and the Economy* (MIT Press 2001).

Shurmer, M, 'Standarisation: A New Challenge for the Intellectual Property System' in Andrew Webster and Kathryn Packer (eds), *Innovation and the Intellectual Property System* (Kluwer Law International 1996).

Sibley, K, 'Disclosure Requirements' in Kenneth Sibley (ed), *The Law and Strategy of Biotechnology Patents* (Butterworth-Heinemann 1994).

Sterckx, S, 'The Ethics of Patenting – Uneasy Justifications' in Peter Drahos (ed), *Death of Patents* (Lawtext Publishing Limited and Queen Mary Intellectual Property Research Institute 2005).

Straus, J, 'Patenting of Human Genes and Living Organisms – The Legal Situation in Europe' in Friedrich Vogel and Reinhard Grunwald (eds), *Patenting of Human Genes and Living Organisms* (Springer 1994).

Straus, J, 'Product Patents on Human DNA Sequences' in F Scott Kieff (ed), *Perspectives on Properties of the Human Genome Project* (Elsevier/Academic Press 2003).

Wadlow, C, 'Utility and Industrial Applicability' in Toshiko Takenaka (ed), *Patent Law and Theory* (Edward Elgar 2008).

Walsh, JP and Wesley M Cohen, 'Research Tool Patenting and Licensing and Biomedical Innovation, in Patents in the Knowledge-Based Economy' in WM Cohen and S Merrill (eds), *Patents in the Knowledge-Based Economy* (National Academies Press 2002).

CONFERENCE PAPERS

David, PA, 'The Evolution of Intellectual Property Institutions and the Panda's Thumb' (Meetings of the International Economic Association, Moscow, August 1992).

Reitzig, M, 'The Private Values of Thicket Patenting: An Empirical Study Across Industries and Patent Types' (DRUID Summer Conference on 'Industrial Dynamics of the New and Old Economy – who is embracing whom?', Copenhagen, June 2002).

JOURNAL ARTICLES

Adelman, DE, 'A Fallacy of the Commons in Biotech Patent Policy' (2005) 20 BTLJ 985.

Aerts, RJ, 'The Industrial Applicability and Utility Requirements for the Patenting of Genomic Inventions: A Comparison between European and US Law' (2004) 26 EIPR 349.

Aerts, RJ, 'Biotechnological Patents in Europe – Functions of Recombinant DNA and Expressed Protein and Satisfaction of the Industrial Applicability Requirement' (2008) 39 IIC 282.

Aljalian, NN, 'The Role of Patent Scope in Biopharmaceutical Patents' (2005) 11 BUJ Sci & Tech L 1.
Andewelt, RB, 'Analysis of Patent Pools under the Antitrust Laws' (1984) 53 Antitrust Law Journal 611.
Andrade, MA and Chris Sander, 'Bioinformatics: From Genome Data to Biological Knowledge' (1997) 8 Current Opinion in Biotechnology 675.
Andrews, LB, 'The Gene Patent Dilemma: Balancing Commercial Incentives with Health Needs' (2002) 2 Hous J Health L & Pol'y 65.
Andrews, LB and Paradise, J, 'Gene Patents: The Need for Bioethics Scrutiny and Legal Change' (2005) 5 Yale J Health Pol'y L & Ethics 403.
Armitage, E, 'Interpretation of European Patents (Art 69 EPC and the Protocol on the Interpretation)' (1983) 14 IIC 811.
Austin, DH, 'Estimating Patent Value and Rivalry Effects: An Event Study of Biotechnology Patents' (1994) Resources for the Future (Washington, DC) Discussion Paper 94-36.
Auth, DR, 'Are ESTs Patentable?' (1997) 15 Nat Biotech 911.
Aymé, S, Gert Matthijs and S Soini, 'Patenting and Licensing in Genetic Testing' (2008) 16 European Journal of Human Genetics (Supp) 405.
Baldock, C and Oliver Kingsbury, 'Where Did It Come From and Where Is It Going? The Biotechnology Directive and Its Relation to the EPC' (2000) 19 Biotechnology Law Report 7.
Baldock, C and others, 'Report Q 150: Patentability Requirements and Scope of Protection of Expressed Sequence Tags (ESTs), Single Nucleotide Polymorphisms (SNPs) and Entire Genomes' (2000) 22 EIPR 39.
Bavec, S, 'Scope of Protection: Comparison of German and English Courts' Case Law' (2004) 8 Marq Intell Prop L Rev 255.
Beck, RL, 'The Prospect Theory of the Patent System and Unproductive Competition' (1983) 5 Research in Law and Economics 193.
Becker, KB, 'Are Natural Gene Sequences Patentable?' (2000) 73 Int Arch Occup Environ Health (Supp) S19.
Beier, F-K, 'The European Patent System' (1981) 14 Vand J Transnat'l L 1.
Bendekgey, L and Diana Hamlet-Cox D, 'Gene Patents and Innovation' (2002) 77 Acad Med 1373.
Benson, JC, 'Resuscitating the Patent Utility Requirement, Again: A Return to *Brenner v. Manson*' (2002) 36 UC Davis Law Review 267.
Bostyn, SJR, 'The Patentability of Genetic Information Carriers' [1999] IPQ 1.
Bostyn, SJR, 'A European Perspective on the Ideal Scope of Protection and the Disclosure Requirement for Biotechnological Inventions in a

Harmonized Patent System: The Quest for the Holy Grail' (2002) 5 J World Intell Prop 1013.
Bozicevic, K, 'Distinguishing Products of Nature from Products Derived from Nature' (1987) 69 JPTOS 415.
Bradshaw, J, 'Gene Patent Policy: Does Issuing Gene Patents Accord with the Purpose of the U.S. Patent System' (2001) 37 Willamette Law Review 637.
Brashear, AD, 'Evolving Biotechnology Patent Laws in the United States and Europe: Are They Inhibiting Disease Research' (2001) 12 Ind Int'l & Comp L Rev 183.
Bryan, E, 'Gene Protection: How Much Is Too Much – Comparing the Scope of Patent Protection for Gene Sequences between the United States and Germany' (2009) 9 J High Tech L 52.
Buchanan, JM and Yong J Yoon, 'Symmetric Tragedies: Commons and Anticommons' (2000) 43 Journal of Law & Economics 1.
Bud, R, 'History of "Biotechnology"' (1989) 337 Nature 10.
Burk, DL and Mark A Lemley, 'Is Patent Law Technology-Specific?' (2002) 17 BTLJ 1155.
Burk, DL and Mark A Lemley, 'Policy Levers in Patent Law' (2003) 89 Virginia Law Review 1575.
Burk, DL and Mark A Lemley, 'Fence Posts or Sign Posts: Rethinking Patent Claim Construction' (2009) 157 U Pa L Review 1743.
Cantor, AE, 'Using the Written Description and Enablement Requirements to Limit Biotechnology Patents' (2000) 14 Harvard Journal of Law & Technology 267.
Carpenter, CC, 'Seeds of Doubt: The European Court of Justice's Decision in *Monsanto v. Cefetra* and the Effect on European Biotechnology Patent Law' [2010] The International Lawyer 1189.
Caulfield, T, E Richard Gold and Mildred K Cho, 'Patenting Human Genetic Material: Refocusing the Debate' (2000) 1 Nature Reviews Genetics 227.
Chang, HF, 'Patent Scope, Antitrust Policy, and Cumulative Innovation' (1995) 26 RAND Journal of Economics 34.
Chavez, MA, 'Gene Patenting: Do the Ends Justify the Means' (2002) 7 Computer L Rev & Tech J 255.
Coase, RH, 'The Problem of Social Cost' (1960) 3 Journal of Law & Economics 1.
Cohen, DL, 'Article 69 and European Patent Integration' (1997) 92 Nw U L Rev 1082.
Cohen, S and others, 'Litigating Biotech Patents in Europe' (2008) 28 IAM Magazine 45.

Cohen, WM, Richard R Nelson and John P Walsh, 'Protecting Their Intellectual Property Assets: Appropriability Conditions and Why the US Manufacturing Firms Patent (or Not)' [2000] NBER Working Paper No 7552.
Collins, KE, 'Propertizing Thought' (2007) 60 SMU L Rev 317.
Conley, JM, 'Gene Patents and the Product of Nature Doctrine' (2009) 84 Chicago-Kent L Rev 109.
Conley, JM and Makowski, R, 'Back to the Future: Rethinking the Product of Nature Doctrine as a Barrier to Biotechnology Patents (Part I)' (2003) 85 JPTOS 301.
Connor, MT, 'European Patents: What's New in 2008 for Applicants' (2008) 90 JPTOS 587.
Cook, T, 'The Human Genome Project: Crucial Questions the Biotechnology Directive Does Not Answer' (2000) 1 European Lawyers 56.
Cook-Deegan, R and Christopher Heaney, 'Patents in Genomics and Human Genetics' (2010) 11 Annual Review of Genomics and Human Genetics 383.
Cornish, WR, 'Scope and Interpretation of Patent Claims under Article 69 of the European Patent Convention' (2000) 4 Int'l Intell Prop L & Pol'y 34.
Correa, CM, 'Efforts to Raise the Bar in Patent Examination Need to be Supported' (2012) 43 IIC 747.
Crespi, RS, 'Biotechnology, Broad Claims and the EPC' (1995) 17 EIPR 267.
Crespi, RS, 'Biotechnology Patenting: The Wicked Animal Must Defend Itself' (1995) 17 EIPR 431.
Crespi, RS, 'Patents on Genes: Can the Issues Be Clarified?' (1999) 3 BIO-Science Law Review 199.
Crespi, RS, 'Patents on Genes: Clarifying the Issues' (2000) 18 Nat Biotech 683.
Davis, PK and others, 'ESTs Stumble at the Utility Threshold' (2005) 23 Nat Biotech 1227.
Demsetz, H, 'Toward a Theory of Property Rights' (1967) 57 Am Econ Rev 354.
Denicolo, V, 'Patent Races and Optimal Patent Breadth and Length' (1996) XLIV The Journal of Industrial Economics 263.
Depoorter, B and Sven Vanneste, 'Putting Humpty Dumpty Back Together: Experimental Evidence of Anticommons Tragedies' (2006) 3 Journal of Law, Economics & Policy 1.
Drahos, P, 'Biotechnology Patents, Markets and Morality' (1999) 21 EIPR 441.

Duffy, JF, 'Harmony and Diversity in Global Patent Law' (2002) 17 BTLJ 685.
Ebersole TJ, C Guthrie and JA Goldstein, 'Patent Pools and Standard Setting in Diagnostic Testing' (2005) 23 Nat Biotech 937.
Eisenberg, RS, 'Patents and the Progress of Science: Exclusive Rights and Experimental Use' (1989) 56 U Chi L Rev 1017.
Eisenberg, RS, 'Intellectual Property Issues in Genomics' (1996) 14 Trends in Biotechnology 302.
Eisenberg, RS, 'Analyze This: A Law and Economics Agenda for the Patent System' (2000) 53 Vand L Rev 2081.
Eisenberg, RS, 'How Can You Patent Genes?' (2002) 2 The American Journal of Bioethics 3.
Eisenberg, RS, 'Why the Gene Patenting Controversy Persists' (2002) 77 Acad Med 1381.
Eisenberg, RS and Merges, RP, 'Opinion Letter as to the Patentability of Certain Inventions Associated with the Identification of Partial cDNA Sequences' (1995) 23 AIPLA Quarterly Journal 1.
Enserink, M, 'Patent Office May Raise the Bar on Gene Claims' (2000) 287 Science 1196.
Epstein, RA and Bruce N Kuhlik, 'Is There a Biomedical Anticommons' (2004) 27 Regulation 54.
Farmer, SJ and Martin Grund, 'Revision of the European Patent Convention & (and) Potential Impact on European Patent Practice' (2008) 36 AIPLA Quarterly Journal 419.
Fitt, R and Edward Nodder, 'Specific, Substantial and Credible? A New Test for Gene Patents' (2006) 9 BIO-Science Law Review 183.
Fitt, R and Edward Nodder, 'The Industrial Applicability of Biotechnology Patents – A New Test for Europe' (2009) 28 Biotechnology Law Report 151.
Franzoni, C and F Lissoni, 'Academic Entrepreneurship, Patents, and Spin-Offs: Critical Issues and Lessons for Europe' [2006] CESPRI.
Gallini, NT, 'Patent Policy and Costly Imitation' [1992] The RAND Journal of Economics 52.
Gallini, NT, 'The Economics of Patents: Lessons from Recent U.S. Patent Reform' (2002) 16 The Journal of Economic Perspectives 131.
Geradin, D, Anne Layne-Farrar and A Jorge Padilla, 'The Complements Problem within Standard Setting: Assessing the Evidence on Royalty Stacking' (2008) 14 BUJ Sci & Tech L 2.
Geuna, A and Federica Rossi, 'Changes to University IPR Regulations in Europe and the Impact on Academic Patenting' (2011) 40 Special Issue: 30 Years After Bayh–Dole – Reassessing Academic Entrepreneurship 1068.

Gielen, C, 'Netherlands: Patents – Biotech Directive – Scope of Protection' (2010) 32 EIPR 93.

Gilbert, R and Carl Shapiro, 'Optimal Patent Length and Breadth' (1990) 21 The RAND Journal of Economics 106.

Gold, ER and Julia Carbone, 'Myriad Genetics: In the Eye of the Policy Storm' (2010) 12 Genetics in Medicine: official journal of the American College of Medical Genetics (Supp) S39.

Golden, JM, 'Biotechnology, Technology Policy, and Patentability: Natural Products and Invention in the American System' (2001) 50 Emory LJ 101.

Goldstein, JA and E Golod, 'Human Genes Patents' (2002) 77 Acad Med 1315.

Grady, MF and Jay I Alexander, 'Patent Law and Rent Dissipation' (1992) 78 Virginia Law Review 305.

Grassler, F and Mary Ann Capria, 'Patent Pooling: Uncorking a Technology Transfer Bottleneck and Creating Value in the Biomedical Research Field' (2003) 9 Journal of Commercial Biotechnology 111.

Greenlee, LL, 'Biotechnology Patent Law: Perspective of the First Seventeen Years, Prospective on the Next Seventeen Years' (1991) 68 Denv UL Rev 127.

Hall, BH and Rosemarie Ham Ziedonis, 'The Patent Paradox Revisited: An Empirical Study of Patenting in the U.S. Semiconductor Industry, 1979–1995' (2001) 32 The RAND Journal of Economics 101.

Hamme Ramirez, H, 'Defending the Privatization of Research Tools: An Examination of the Tragedy of the Anticommons in Biotechnology Research and Development' (2004) 53 Emory Law Journal 359.

Hardin, G, 'The Tragedy of the Commons' (1968) 162 Science 1243.

Hasan, SA, 'A Call for Reconsideration of the Strict Utility Standard in Chemical Patent Practice' (1994) 9 High Technology Law Journal 245.

Heller, MA, 'The Tragedy of the Anticommons: Property in the Transition from Marx to Markets' (1998) 111 Harvard Law Review 621.

Heller, MA and Eisenberg, RS, 'Can Patents Deter Innovation? The Anticommons in Biomedical Research' (1998) 280 Science 698.

Henderson, R, Adam B Jaffe and Manuel Trajtenberg, 'Universities as a Source of Commercial Technology: A Detailed Analysis of University Patenting, 1965–1988' (1998) 80 Review of Economics and Statistics 119.

Herdegen, M, 'Patenting Human Genes and Other Parts of the Human Body under EC Biotechnology Directive' (2001) 4 BIO-Science Law Review 102.

Hettinger, N, 'Patenting Life: Biotechnology, Intellectual Property, and Environmental Ethics' (1994) 22 BC Envtl Aff L Rev 267.

Holman, CM, 'Patent Border Wars: Defining the Boundary between Scientific Discoveries and Patentable Inventions' (2007) 25 TRENDS in Biotechnology 539.

Holmes, S, 'The New Biotech Patents Directive' [1998] Patent World 16.

Howlett, MJ and Andrew F Christie, 'An Analysis of the Approach of the European, Japanese and United States Patent Offices to Patenting Partial DNA Sequences (ESTs)' (2003) 34 International Review of Industrial Property and Copyright 581.

Jacob, M, 'Patentability of Natural Products' (1970) 52 J Pat Off Soc'y 473.

Jameson, SA, 'A Comparison of the Patentability and Patent Scope of Biotechnological Inventions in the United States and the European Union' (2007) 35 AIPLA Quarterly Journal 193.

Janis, MD, 'On Courts Herding Cats: Contending with the Written Description Requirement (and Other Unruly Patent Disclosure Doctrines)' (2000) 2 Washington University Journal of Law & Policy 55.

Karczewski, LA, 'Biotechnological Gene Patent Applications: The Implications of the USPTO Written Description Requirement Guidelines on the Biotechnology Industry' (1999) 31 McGeorge Law Review 1043.

Kelmelyte, I, 'Can Living Things Be Objects of Patents?' (2005) 2 International Journal of Baltic Law 1.

Kieff, FS, 'Facilitating Scientific Research: Intellectual Property Rights and the Norms of Science – A Response to Rai and Eisenberg' [2000] Law and Economics Papers, Working Paper 43.

Kiley, TD, 'Patents on Random Complementary DNA Fragments?' (1992) 257 Science 915.

Kitch, EW, 'The Nature and Function of the Patent System' (1977) 20 Journal of Law & Economics 265.

Klemperer, P, 'How Broad Should the Scope of Patent Protection Be?' (1990) 21 The RAND Journal of Economics 113.

Klofsten, M and D Jones-Evans, 'Comparing Academic Entrepreneurship in Europe –The Case of Sweden and Ireland' (2000) 14 Small Business Economics 299.

Ko, Y, 'An Economic Analysis of Biotechnology Patent Protection' (1992) 102 The Yale Law Journal 777.

Kock, MA, 'Purpose-Bound Protection for DNA Sequences: In through the Back Door?' (2010) 5 JIPLP 495.

Kowalski, TJ, A Maschio and SH Megerditchian, 'Dominating Global Intellectual Property: Overview of Patentability in the USA, Europe and Japan' (2003) 9 Journal of Commercial Biotechnology 305.

Krauss, JB and Toshiko Takenaka, 'A Special Rule for Compound Protection for DNA-Sequences – Impact of the ECJ "Monsanto" Decision on Patent Practice' (2001) 93 JPTOS 189.

Laakmann, AB, 'An Explicit Policy Lever for Patent Scope' (2012) 19 Mich Telecomm & Tech L Rev 43.

Lacy, PA, 'Gene Patenting: Universal Heritage vs. Reward for Human Effort' (1998) 77 Or L Rev 783.

Lampe, RL and Petra Moser, 'Do Patent Pools Encourage Innovation? Evidence from the 19th-Century Sewing Machine Industry' (2009) National Bureau of Economic Research, Working Paper 15061.

Lander, ES and others, 'Initial Sequencing and Analysis of the Human Genome' (2001) 409 Nature 860.

Lawrence, S, 'US Court Case to Define EST Patentability' (2005) 23 Nat Biotech 513.

Lech, KF, 'Human Genes without Functions: Biotechnology Tests the Patent Utility Standard' (1993) 27 Suffolk UL Rev 1631.

Lemley, MA, 'Software Patents and the Return of Functional Claiming' [2012] Stanford Public Law Working Paper No 2117302.

Lemley, MA and Shapiro, C, 'Patent Holdup and Royalty Stacking' (2007) 85 Texas Law Review 1991.

Lenoir, N, 'Patentability of Life and Ethics' (2003) 326 Comptes Rendus Biologies 1127.

Lerner, J, 'The Importance of Patent Scope: An Empirical Analysis' (1994) 25 The RAND Journal of Economics 319.

Leung, S, 'The Commons and Anticommons in Intellectual Property' [2011] UCL Jurisprudence Review 16.

Leverve, F, 'Patents on Genes, Usefulness, and the Requirement of Industrial Application' [2009] JIPLP 289.

Levin, RC and others, 'Appropriating the Returns from Industrial Research and Development' [1987] 3 Brookings Papers on Economic Activity (Special Issue On Microeconomics) 783.

Lissoni, F and others, 'Academic Patenting in Europe: New Evidence from the KEINS Database' (2008) 16 Research Evaluation 87.

Lissoni, F and others, 'Academic Patenting and the Professor's Privilege: Evidence on Denmark from the KEINS Database' (2009) 36 Science and Public Policy 595.

Llewelyn, M, 'Industrial Applicability/Utility and Genetic Engineering: Current Practices in Europe and the United States' (1994) 16 EIPR 473.

Lockhart, DJ and Elisabeth A Winzeler, 'Genomics, Gene Expression and DNA Arrays' (2000) 405 Nature 827.

Lopez-Beverage, CD, 'Should Congress Do Something about Upstream Clogging Caused by the Deficient Utility of Expressed Sequence Tag Patents' (2005) 10 J Tech L & Pol'y 35.

Luukkonen, M, 'Gene Patents: How Useful Are the New Utility Requirements' (2000) 23 Thomas Jefferson Law Review 337.

Machin, N, 'Prospective Utility: A New Interpretation of the Utility Requirement of Section 101 of the Patent Act' (1999) 87 California Law Review 421.

Machlup, F and Edith Penrose, 'The Patent Controversy in the Nineteenth Century' (1950) 10 The Journal of Economic History 1.

Maebius, SB, 'Novel DNA Sequences and the Utility Requirement: The Human Genome Initiative' (1992) 74 JPTOS 651.

Mazzoleni, R and Richard R Nelson, 'The Benefits and Costs of Strong Patent Protection: A Contribution to the Current Debate' (1998) 27 Research Policy 273.

McBride, MS, 'Patentability of Human Genes: Our Patent System Can Address the Issues without Modification' (2001) 85 Marq L Rev 511.

McCoy, J, 'Patenting Life in the European Community: The Proposed Directive on the Legal Protection for Biotechnological Inventions' (1993) 4 Fordham Intell Prop Media & Ent LJ 501.

McLeod, BW, 'The "Real World" Utility of miRNA Patents: Lessons Learned from Expressed Sequence Tags' (2011) 29 Nat Biotech 129.

Meigs, JT, 'Biotechnology Patent Prosecution in View of PTO's Utility Examination Guidelines' (2001) 83 JPTOS 451.

Merges, RP and Richard R Nelson, 'On The Complex Economics of Patent Scope' (1990) 90 Columbia Law Review 839.

Merges, RP and Richard R Nelson, 'On Limiting or Encouraging Rivalry in Technical Progress: The Effect of Patent Scope Decisions' (1994) 25 Journal of Economic Behavior & Organization 1.

Michaels, CA, 'Biotechnology and the Requirement for Utility in Patent Law' (1994) 76 JPTOS 247.

Minssen, T and David Nilsson, 'The Industrial Application Requirement for Biotech Inventions in Light of Recent EPO & UK Case Law: A Plausible Approach or a Mere "Hunting License"?' [2012] EIPR 689.

Mirabel, EP, 'Practical Utility Is a Useless Concept' (1986) 36 Am UL Rev 811.

Mireles, MS, 'An Examination of Patents, Licensing, Research Tools, and the Tragedy of the Anticommons in Biotechnology Innovation' (2004) 38 University of Michigan Journal of Law Reform 141.

Mohan-Ram, V, Richard Peet and Philippe Vlaemminck, 'Biotech Patent Infringement in Europe: The "Functionality" Gatekeeper' (2011) 10 J The John Marshall Review of Intellectual Property 1540.

Moore, S, 'Challenge to the Biotechnology Directive' (2002) 24 EIPR 149.

Morgan, G, Royle M and Cohen S, 'Cargill vs. Monsanto' (2008) 27 Biotechnology Law Report 109.

Morgan, G and Lisa A Haile, 'A Shadow Falls over Gene Patents in the United States and Europe' (2010) 28 Nat Biotech 1172.

Mowery, D and A Ziedonis 'Academic Patent Quality and Quantity Before and After the Bayh–Dole Act in the United States' (2002) 31 Research Policy 399.

Mowery, D and others, 'The Growth of Patenting and Licensing by U.S. Universities: An Assessment of the Effects of the Bayh–Dole Act of 1980' (2001) 30 Research Policy 99.

Muller, FE and Harold C Wegner, 'The 1976 German Patent Law' (1977) 59 J Pat Off Soc'y 91.

Murray, F and Scott Stern, 'When Ideas Are Not Free: The Impact of Patents on Scientific Research' (2006) 7 Innovation Policy and the Economy 33.

Murray, K and Esther van Zimmeren, 'Dynamic Patent Governance in Europe and the United States: The Myriad Example' (2011) 19 Cardozo J Int'l & Comp L 287.

Newberg, JA, 'Antitrust, Patent Pools, and the Management of Uncertainty' (2000) 3 Atlantic Law Journal 1.

Nordhaus, W, 'The Optimum Life of a Patent: Reply' (1972) 62 American Economic Review 428.

O'Donoghue, T, Suzanne Scotchmer and Jacques-François Thisse, 'Patent Breadth, Patent Life and the Pace of Technological Progress' (1998) 7 Journal of Economics & Management Strategy 3.

Odell-West, A, '"Gene"-Uinely Patentable? The Distinction in Biotechnology between Discovery and Invention in US and EU Patent Law' [2011] IPQ 304.

Odell-West, A, 'Has the Commodore Steered the Fleet onto the Rocks? Biotechnology and the Requirement for Industrial Applicability' (2013) 4 IPQ 279.

Oser, A, 'Patenting (Partial) Gene Sequences Taking Particular Account of the EST Issue' (1999) 30 IIC 1.

Palombi, L, 'Patentable Subject Matter, TRIPS and the European Biotechnology Directive: Australia and Patenting Human Genes' (2003) 26 UNSWLJ 782.

Palombi, L, 'The Impact of TRIPS on the Validity of the European Biotechnology Directive' [2005] JIBL 62.

Parisi, F, Norbert Schulz and Ben Depoorter, 'Duality in Property: Commons and Anticommons' (2005) 25 International Review of Law and Economics 578.

Pechhold, AK, 'The Evolution of the Doctrine of Equivalents in the United States, United Kingdom, and Germany' (2005) 87 JPTOS 411.

Phillips, J, 'The English Patent as a Reward for Invention: The Importation of an Idea' (1982) 3 The Journal of Legal History 71.

Pila, J, 'Article 52(2) of the Convention for the Grant of European Patents: What Did the Framers Intend? A Study of the Travaux Preparatoires' (2005) 36 IIC 755.
Pila, J, 'On the European Requirement for an Invention' (2010) 41 IIC 906.
Pippen, SC, 'Dollars and Lives: Finding Balance in the Patent Gene Utility Doctrine' (2006) 12 BUJ Sci & Tech L 193.
Pires de Carvalho, N, 'The Problem of Gene Patents' (2004) 3 Wash U Global Stud L Rev 701.
Pisano, GP, 'Can Science Be a Business? Lessons from Biotech' (2006) 84 Harvard Business Review 114.
Rai, AK, 'Fostering Cumulative Innovation in the Biopharmaceutical Industry: The Role of Patents and Antitrust' (2001) 16 Berkeley Tech LJ 813.
Resnik, DB, 'A Biotechnology Patent Pool: An Idea Whose Time Has Come?' (2003) 3 The Journal of Philosophy, Science & Law.
Restaino, LG, Steven E Halpern and Eric L Tang, 'Patenting DNA-Related Inventions in the European Union, United States and Japan: A Trilateral Approach or a Study in Contrast?' (2003) 2003 UCLA JL & Tech 2.
Roberts, T, 'Broad Claims for Biotechnological Inventions' (1995) 17 EIPR 267.
Scalise, DG and Daniel Nugent, 'Patenting Living Matter in the European Community: Diriment of the Draft Directive' (1992) 16 Fordham Int'l LJ 990.
Scherer, FM, 'Nordhaus' Theory of Optimal Patent Life: A Geometric Reinterpretation' (1972) 62 American Economic Review 422.
Schertenleib, D, 'The Patentability and Protection of DNA-Based Inventions in the EPO and the European Union' (2003) 25 EIPR 125.
Schertenleib, D, 'The Patentability and Protection of Living Organisms in the European Union' (2004) 26 EIPR 203.
Schuster, MI, 'Sufficient Disclosure in Europe: Is There a Separate Written Description Doctrine under the European Patent Convention' (2007) 76 UMKC L Rev 491.
Scotchmer, S, 'Standing on the Shoulders of Giants: Cumulative Research and the Patent Law' (1991) 5 Journal of Economic Perspectives 30.
Scott-Ram, N, 'Biotechnology Patenting in Europe: The Directive on the Legal Protection of Biotechnological Inventions: Is This the Beginning or the End?' [1998] BIO-Science Law Review 43.
Sena, G, 'Directive on Biotechnological Inventions: Patentability of Discoveries' (1999) 30 IIC 731.

Sharples, A, 'Industrial Applicability for Genetics Patents – Divergences between the EPO and the UK' (2011) 33 EIPR 72.
Sharples, A and Smith, RL, 'Patents Claiming Genetic Sequences' (2009) 216 Patent World 32.
Sherman, B, 'Patent Claim Interpretation: The Impact of the Protocol on Interpretation' (1991) 54 MLR 499.
Simon, JHM and others, 'Managing Severe Acute Respiratory Syndrome (SARS) Intellectual Property Rights: The Possible Role of Patent Pooling' (2005) 83 Bulletin of the World Health Organization (WHO) 707.
Smith Hughes, S, 'Making Dollars out of DNA: The First Major Patent in Biotechnology and the Commercialization of Molecular Biology, 1974–1980' (2001) 92 Isis 541.
Soobert, AM, 'Analyzing Infringement by Equivalents: A Proposal to Focus the Scope of International Patent Protection' (1996) 22 Rutgers Computer & Tech L J 189.
Spencer, M, 'What's the Use?!' (2009) 10 BIO-Science Law Review 56.
Stephan, PE and Stephen S Everhart, 'The Changing Rewards to Science: The Case of Biotechnology' (1998) 10 Small Business Economics 141.
Straus, J, 'Product Patents on Human DNA Sequences: An Obstacle for Implementing the EU Biotech Directive?' (2003) 50 Advances in Genetics 65.
Summers, TM, 'The Scope of Utility in the Twenty-First Century: New Guidance for Gene-Related Patents' (2002) 91 Georgetown Law Journal 475.
Taylor, MR and Jerry Cayford, 'The U.S. Patent System and Developing Country Access to Biotechnology: Does the Balance Need Adjusting?' [2002] Resources for the Future 2.
Thambisetty, S, 'Legal Transplants in Patent Law: Why Utility is the New Industrial Applicability' Law, Society and Economy Working Papers 6/2008 (2009) 49 Jurimetrics J 155.
Thumm, N, 'Patents for Genetic Inventions: A Tool to Promote Technological Advance or a Limitation for Upstream Inventions?' (2005) 25 Technovation 1410.
Tischer, E and others, 'The Human Gene for Vascular Endothelial Growth Factor. Multiple Protein Forms Are Encoded through Alternative Exon Splicing' (1991) 266 Journal of Biological Chemistry 11947.
Van Caenegem, W, 'The Technicality Requirement, Patent Scope and Patentable Subject Matter in Australia' (2002) 13 AIPJ 309.
Van Overwalle, G, 'Editorial, The CJEU's Monsanto Soybean Decision and Patent Scope – as Clear as Mud?' (2011) 42 IIC.

Van Overwalle, G and others, 'Models for Facilitating Access to Patents on Genetic Inventions' (2006) 7 Nat Rev Genet 143.
Van Zeebroeck, N, Bruno van Pottelsberghe de la Potterie and Dominique Guellec, 'Patents and Academic Research: A State of the Art' (2008) 9 Journal of Intellectual Capital 246.
Van Zimmeren, E and Geertrui Van Overwalle, 'A Paper Tiger? Compulsory License Regimes for Public Health in Europe' (2011) 42 IIC 4.
Venter, JC and others, 'The Sequence of the Human Genome' (2001) 291 Science 1304.
Verbeure, B and others, 'Patent Pools and Diagnostic Testing' (2006) 24 TRENDS in Biotechnology 115.
Watson, JD and FHC Crick, 'Genetical Implications of the Structure of Deoxyribonucleic Acid' (1953) 171 Nature 964.
Watson, JD and FHC Crick, 'Molecular Structure of Nucleic Acids: A Structure for Deoxyribose Nucleic Acid' (1953) 171 Nature 737.
Webber, P, 'Limitation of Gene Patents to Functioning DNA' (2011) 17 Journal of Commercial Biotechnology 201.
Wee Loon, N-L, 'Patenting of Genes – A Closer Look at the Concepts of Utility and Industrial Applicability' [2003] IIC 393.
Weekes, RN, 'Challenging the Biotechnology Directive' [2004] European Business Law Review 325.
Woessner, WD, 'The Evolution of Patents on Life – Transgenic Animals, Clones and Stem Cells' (2001) 83 JPTOS 830.
Zech, H, 'Nanotechnology – New Challenges for Patent Law?' (2009) 6 SCRIPT-ed 147.
Zech, H and Ensthaler, J, 'Stoffschutz Bei Gentechnischen Patenten, Gewerblicher Rechtsschutz Und Urheberrecht' [2006] GRUR 529.
Zimmer, F-J and Svenja Sethmann, 'Act Implementing the Directive on the Legal Protection of Biotechnological Inventions in Germany (BioPatG)' (2005) 24 Biotechnology Law Report 561.
Zuhn, DL, 'DNA Patentability: Shutting the Door to the Utility Requirement' (2000) 34 J Marshall L Rev 973.

ONLINE SOURCES

BIO, 'BIO's testimony on patent reform to maximize innovation in the biotechnology industry' (25 March 1999) <http://www.bio.org/advocacy/letters/bios-testimony-patent-reform-maximize-innovation-biotechnology-industry> accessed 4 September 2014.
BIO, 'Statement on US Supreme Court review of isolated DNA patents' (13 June 2013) <http://www.bio.org/media/press-release/statement-us-supreme-court-review-isolated-dna-patents> accessed 5 August 2014.

CIPA, 'Patenting of human gene sequences' (April 1994 (last revised June 2006)) <http://www.cipa.org.uk/pages/info-papers-human> accessed 27 July 2013.

Cole, S, 'Virtual friend fires employee' (Naked Law, 1 May 2009) <http://www.nakedlaw.com/2009/05/index.html> accessed 19 November 2009.

EPO, 'Quality over quantity: on course to raise the bar' (2008) <http://www.epo.org/about-us/office/annual-report/2008/focus.html> accessed 10 October 2012.

Marchal, V, 'Brevets, marques, dessins et modèles. Évolution des protections de propriété industrielle au XIXe siècle en France', Documents pour l'histoire des techniques (17, 1er semestre 2009, L'invention technique et les figures de l'inventeur (XVIIIe–XXe siècles)) (2009) <http://dht.revues.org/392> accessed 1 March 2015.

Syngenta AG, Syngenta AG's e-licensing platform TraitAbility <http://www3.syngenta.com/global/e-licensing/en/e-licensing/About/Pages/About.aspx> accessed 20 February 2015.

Turner, JL, 'Patent thickets, trolls and unproductive entrepreneurship' (2011) <http://ssrn.com/abstract=1916798> accessed 19 September 2012.

UK BioIndustry Association 'Letter to the Guardian regarding gene patenting' (23 November 2000) <http://www.bioindustry.org/newsandresources/bia-news/letter-to-the-guardian-regarding-gene-patenting/> accessed 13 February 2013.

PATENT APPLICATIONS

EPO, European Patent EP0041313, DNA sequences, recombinant DNA molecules and processes for producing human fibroblast interferon-like polypeptides (1981).

EPO, European Patent EP0101309, Molecular cloning and characterization of a gene sequence coding for human relaxin (1983).

EPO, European Patent EP0411946, DNA encoding human GP130 protein (1990).

EPO, European Patent EP0431065, A full length cDNA encoding a human laminin binding protein (1989).

EPO, European Patent EP0546090, Glyphosate tolerant 5-enolpyruvylshikimate-3-phosphate synthases (1991).

EPO, European Patent EP0630405, Novel V28 seven transmembrane receptor (1993).

EPO, European Patent EP0939804, Neutrokine alpha (1996).

UKIPO, Patent BL O/286/05, Aeomica Inc (2005).

USPTO, Patent Application No 07/716831, Sequences (applied 20 June 1991 by J Craig Venter and Mark Adams).

REPORTS AND OFFICIAL DOCUMENTS

Council of Europe, Consultative Assembly Documents, 1 Session (1949), Council of Europe Doc No 75.

Council of Europe, *Questionnaire Drawn Up by the Bureau of the Committee of Experts on Patents*, CM/12 (52) 149 (17 November 1952).

Council of Europe, *Comparative Study of Substantive Law in Force in the Countries Represented on the Committee of Experts on Patents Presented by the Secretariat-General Mr Gajac*, EXP/Brev (53) 18 (7 November 1953).

Council of Europe, *Memorandum on the Unification of Legislation (Item 4 of the Agenda for the Meeting of 28 November 1960) by Rapporteur-General Mr. Finniss*, EXP/Brev (60) 7 (28 November 1960).

Council of Europe, *First Preliminary Draft of a European Patent Law Convention* (14 March 1961).

Council of Europe, Committee of Experts, *Report of the Meeting Held in Paris on 16 and 17 March 1961*, EXP/Brev B (61) 3 (24 March 1961).

Council of Europe, Proceedings of the 1st meeting of the PWP held at Brussels from 17 to 28 April 1961, Council of Europe Doc IV/2767/61-E (1961).

Council of Europe, EXP/Brev/Misc (61) 7 (2 May1961).

Council of Europe, Committee of Experts, *Report of the Committee of Experts to the Committee of Ministers on the Meeting held at Strasbourg from 10th to 13th July of 1962*, EXP/Brev (61) 8 (13 December 1961).

Council of Europe, *Second Preliminary Draft of a Convention in a European System for the Grant of Patents* (March 1972).

Council of Europe, Council Common Position of 26 February 1998 [1998] OJ C 110/17.

Council of Europe, Presidency Conclusions, Lisbon European Council, 23–24 March 2000.

Council of Europe, Presidency Conclusions, Stockholm European Council, 23–24 March 2001.

Cowin, R and others, 'Policy Options for the Improvement of the European Patent System' (2007) Scientific Technology Options Assessment (STOA) of the European Parliament.

EC, Communication to the Council 'Biotechnology in the Community' [COM(83) 672 final/2] – Annex of October 1983.

EC, White Paper from the Commission to the European Council on Completing the Internal Market of 14 June 1985 [COM(85) 310 final].

EC, Commission of the European Communities Proposal for a Council Directive on the Legal Protection of Biotechnological Inventions (1988) [COM(88) 496 final – SYN 159] OJ C10/3.

EC, Amended Proposal for a Council Directive on the Legal Protection of Biotechnological Inventions (1992) [COM(92) 589 final – SYN 159] OJ C44/36.

EC, Commission Proposal for a European Parliament and Council Directive on the Legal Protection of Biotechnological Inventions (1996) [COM(95) 661 final] OJ C296/4.

EC, Amended Proposal for a European Parliament and Council Directive on the Legal Protection of Biotechnological Inventions (1997) [COM(97) 446 final].

EC, Communication from the Commission to the Council, the European Parliament, the Economic and Social Committee and the Committee of the Regions of 23 January 2002: 'Life Sciences and Biotechnology – A Strategy for Europe' (2002) [COM(2002) 27 final] OJ C55/3.

EC, Report from the Commission to the European Parliament and the Council – Development and Implications of Patent Law in the Field of Biotechnology and Genetic Engineering of 7 October 2002 [COM(2002) 545 final].

EC, *Facing the Challenge: The Lisbon Strategy for Growth and Employment – Report from the High Level Group chaired by Wim Kok* (European Communities 2004).

EC, Report from the Commission to the European Parliament, the Council, the Committee of the Regions and the European Economic and Social Committee of 29 June 2005 'Life Sciences and Biotechnology – A Strategy for Europe' Third progress report and future orientations [COM(2005) 286 final].

EC, Report from the Commission to the Council and the European Parliament – Development and Implications of Patent Law in the Field of Biotechnology and Genetic Engineering (SEC(2005) 943) of 14 July 2005 [COM(2005) 312 final].

EC, Communication from the Commission to the Council, the European Parliament, the European Economic and Social Committee and the Committee of the Regions on the Mid-Term Review of the Strategy on Life Sciences and Biotechnology [COM(2007) 175 final].

EPO, Decision of the Administrative Council of 16 June 1999 Amending the Implementing Regulations to the European Patent Convention (OJ EPO 1999, 437).

EPO, Guidelines for Examination in the European Patent Office (2002).

EPO, Decision of the Administrative Council of the European Patent Organisation of 9 December 2004 Amending the Implementing Regulations to the European Patent Convention and the Rules Relating to Fees (OJ EPO 2004, 11).

EPO, Annual Report 2011.

EPO, Decision of the President of the European Patent Office dated 28 April 2011 on the Filing of Sequence Listings (OJ EPO 2011, 372).

EPO, Case Law of the Boards of Appeal of the European Patent Office, Seventh Edition (September 2013).

EPO, Notice from the European Patent Office Dated 18 October 2013 Concerning the Filing of Sequence Listings (OJ EPO 2013, 542).

EPO, Guidelines for Examination in the European Patent Office (2014).

EPO and JPO and USPTO, Trilateral Project B3b, Comparative study on biotechnology patent practices – Theme: Patentability of DNA fragments (Trilateral offices 2000).

EPO and JPO and USPTO, Trilateral Project B3b, Mutual understanding in search and examination – Comparative Study on Biotechnology Patent Practices – Theme: Nucleic acid molecule-related inventions whose functions are inferred based on homology search (Trilateral offices 2000).

European Parliament, Decision on the Joint Text Approved by the Conciliation Committee for a European Parliament and Council Directive on the Legal Protection of Biotechnological Inventions [1995] OJ C68/26.

European Parliament, Report on the Proposal for a European Parliament and Council Directive on the Legal Protection of Biotechnological Inventions (1997) [COM(95)0661 – C4-0063/96 – 95/0350(COD)].

European Parliament, Decision of the European Parliament of 12 May 1998 [1998] OJ C167/5.

European Parliament, Resolution on Patents for Biotechnological Inventions of 26 October 2005 [2006] OJ C 272E/440.

European Parliament, Resolution of 20 October 2010 on the Financial, Economic and Social Crisis: Recommendations Concerning Measures and Initiatives to be Taken (mid-term report) (2009/2182 (INI)) [2012] OJ C 70E/19.

Gowers, A, *Gowers Review of Intellectual Property* (Independent report, UK Government 2006).

Machlup, F, 'An Economic Review of the Patent System' (1958) Study No 15 of the Subcommittee of Patents, Trademarks and Copyrights of the Committee on the Judiciary – United States Senate 85th Congress, Second Session.

NAPAG, *Intellectual Property & the Academic Community* (The Royal Society 1995).

NIH, Report of the National Institutes of Health Working Group on Research Tools (NIH 1998).

Nuffield Council on Bioethics, *The Ethics of Patenting DNA* (Nuffield Council on Bioethics 2002).

OECD, 'Genetic Inventions, Intellectual Property Rights and Licensing Practices, Report of a Workshop Organized by the OECD Working Party on Biotechnology' (OECD 2002).

OECD, Compendium of Patent Statistics (2008).

Royal Society Working Group on Intellectual Property, *Keeping Science Open: The Effect of Intellectual Property Policies on the Conduct of Science* (The Royal Society 2003).

Swiss Federal Institute of Intellectual Property, 'Research and Patenting in Biotechnology: A Survey in Switzerland' (Swiss Federal Institute of Intellectual Property 2003).

UKIPO, Lambert Review of Business-University Collaboration (Final report, UKIPO 2003).

UKIPO, Lambert Toolkit (UKIPO 2005).

UKIPO, 'Patent Thickets: An Overview' (UKIPO 2011).

UKIPO, 'Intellectual Asset Management for Universities' (UKIPO 2014).

UKIPO, Manual of Patent Practice (UKIPO 2014).

UN Conference on Environment and Development, Rio de Janeiro, 3–14 June 1992 – the Earth Summit Agreements: Agenda 21, the Rio Declaration on Environment and Development, the Statement of Forest Principles, the UN Framework Convention on Climate Change and the UN Convention on Biological Diversity.

USPTO, 'Patent Pools: A Solution to the Problem of Access in Biotechnology Patents?' (USPTO 2000).

USPTO, Utility Examination Guidelines (2001).

WIPO, Standing Committee on the Law of Patents, *'Industrial Applicability' and 'Utility' Requirements: Commonalities and Differences* (Ninth Session, Geneva 12–16 May 2003).

UNPUBLISHED THESES

Clarkson, G, 'Objective Identification of Patent Thickets: A Network Analytic Approach' (DPhil thesis, Cambridge, MA: Harvard University 2004).

Zatorski, D, 'The Tragedy of the Anticommons in Biotechnology' (DPhil thesis, The Jagiellonian University in Krakow Intellectual Property Law Institute 2011).

OTHER DOCUMENTS

HUGO, HUGO Intellectual Property Committee Statement on Patenting Issues Related to Early Release of Raw Sequence Data (May 1997).

HUGO, HUGO Intellectual Property Committee Statement on Patenting of DNA Sequences in Particular Response to the European Biotechnology Directive (April 2000).

WIPO, WIPO National Seminar on Intellectual Property, *The International Protection of Industrial Property: From the Paris Convention to the TRIPs Agreement* (Lecture prepared by Professor Michael Blakeney, Cairo 17–19 February 2003).

Index